Student Study (

to accompany

Foundations in Microbiology

Sixth Edition

Kathleen Park Talaro
Pasadena City College

Prepared by
Nancy Boury
Iowa State University

Boston Burr Ridge, IL Dubuque, IA New York San Francisco St. Louis
Bangkok Bogotá Caracas Kuala Lumpur Lisbon London Madrid Mexico City
Milan Montreal New Delhi Santiago Seoul Singapore Sydney Taipei Toronto

The McGraw-Hill Companies

Student Study Guide to accompany
FOUNDATIONS IN MICROBIOLOGY, SIXTH EDITION
KATHLEEN PARK TALARO

Published by McGraw-Hill Higher Education, an imprint of The McGraw-Hill Companies, Inc., 1221 Avenue of the Americas, New York, NY 10020.

2 3 4 5 6 7 8 9 0 QPD/QPD 0 9 8 7

ISBN: 978-0-07-299494-0
MHID: 0-07-299494-0

www.mhhe.com

Introduction

Welcome to the wonderful world of microbiology! You are well on your way to learning more about the millions of microscopic inhabitants with which we share this planet. The *Foundations of Microbiology* text is a great place to start gathering more information. This study guide was written with students in mind as a tool to help focus your efforts and make the most of your study time.

Microbiology, like any discipline, has a vocabulary all its own. It has been said that there are more terms in the average beginning biology class than in a first-semester foreign language class. In order to communicate microbiological information effectively with instructors, classmates, and future colleagues, you must first learn the language. One of the best ways to learn this new language is to increase your exposure to it. Attend class, read your text, and form study groups. With this study guide, answer all the concept questions in your own terms first, then overlay the specialized terminology of the microbiologist.

Knowledge of terminology is very important, but not adequate as an end in itself. College-level courses will require you to possess higher-order thinking skills. Comprehension and application skills will enable you to move beyond the *memorize-and-regurgitate* methods of studying that you may be using. If you can apply the material you are studying you will be able to compare and contrast related and unrelated concepts. You will also be able to generalize trends based on specific examples, or make educated guesses about specific examples based on known trends. These are the skills that allow you to answer "what if" questions and to troubleshoot problems. These are also the skills that will help you succeed both in class and out.

I have written this study guide with both your, the students', and my fellow instructors' expectations in mind. I have taught both entry-level and senior-level science courses at the college and university settings for several years. Each year I prepare study guides based on the text I'm using to help my students focus their efforts and make more efficient use of their study time. With this edition I have divided each chapter into five parts, with a key provided at the end of the book.

I. Building Your Knowledge
- *Highlights key concepts*
- *Reviews material from the text*
- *Provides an opportunity to use the terminology discussed in the text*

II. Organizing Your Knowledge
- *Provides structure and a framework for the material*
- *Reviews the basics of each chapter*
- *Additional practice with using the vocabulary of microbiology*
- *Compares and contrasts several related topics*

III. Self-Test: Vocabulary
- *An opportunity to review basic terms introduced in each chapter*
- *Formatted as matching or crossword puzzles (for fun)*

IV. Self-Test: Multiple Choice
- *Gives sample exam questions and tests your preparedness*
- *Most helpful if completed without notes or text to reference*

V. Applying your Knowledge
- *Applying material from each chapter relevant to real-world problems*

I would like to thank the reviewers of the previous editions of the study guide—Daniel K. Brannan (Abilene Christian University), Elizabeth Godrick (Boston University), and Julie Huggins (Arkansas State University); their suggestions were very helpful. Thanks to Darlene Schueller for her guidance and patience as I've worked my way through this project. I would also like to thank the students in my first-year and senior microbiology courses for their suggestions and even their constructive criticism. I will leave you with the advice I give to my students each semester, "Your study guide should be your best friend, not a passing acquaintance." Enjoy!

Warm regards,
Dr. Nancy Boury

Table of Contents

Chapter 1 The Main Themes of Microbiology

Building Your Knowledge

1. The science of microbiology has several branches. Describe three separate branches of microbiology.

2. Why are most microbes both easier and more difficult to study than larger organisms?

3. Fill in the following events from the figure below: appearance of eucaryotes, Earth's origins, appearance of procaryotes, appearance of humans, reptiles, mammals, and cockroaches/termites.

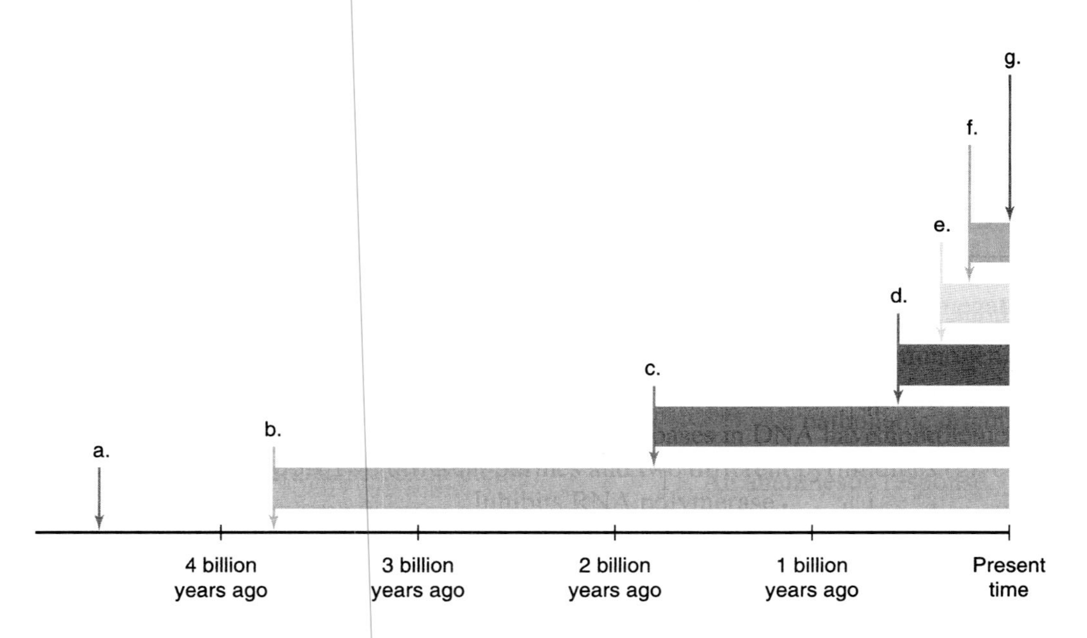

4. Where have scientists found microbes?

   Are procaryotes or eucaryotes the most numerous organisms on the planet?

5. How are photosynthesis and decomposition related to one another?

6. What is genetic engineering? Name several compounds that microbes have been used to make.

7. Bioremediation utilizes microbes to do what task?

8. What are pathogens?

9. How are procaryotes and eucaryotes similar to one another, but different from viruses?

10. How are procaryotes and eucaryotes different from one another?

11. Several people have had a pivotal role in the development of the modern science of microbiology. What was the contribution of each of the following individuals?

    Antonie van Leeuwenhoek

    Louis Pasteur

    Robert Koch

    Joseph Lister

12. What are the three domains of life? Which are procaryotic?

13. What is the correct way to write MYCOBACTERIA LEPRAE?

Self-Test: Vocabulary

1. ____ found everywhere	A. Archaea
2. ____ microbe made up of protein and nucleic acid	B. aseptic techniques
3. ____ use of light energy to fix carbon from carbon dioxide to sugar	C. binomial nomenclature
4. ____ breakdown of organic molecules to simpler forms	D. bioremediation
5. ____ using microbes to produce drugs, foods, and industrial products	E. biotechnology
6. ____ use of microbes to clean up toxic wastes	F. decomposition
7. ____ disease-causing agents	G. hypothesis
8. ____ the concept that living things can arise from non-living things	H. pathogens
9. ____ testable, tentative explanation of an observation	I. photosynthesis
10. ____ procedures used to limit the growth of microbes, particularly in medical settings	J. spontaneous generation
11. ____ hierarchical scheme for classifying organisms	K. taxonomy
12. ____ naming of organisms with two names (genus and species)	L. ubiquitous
13. ____ simple procaryotes that often live in extreme environments	M. virus

Self-Test: Multiple Choice

1. ___________ monitor and try to control the spread of diseases in communities.
   a. Genetic engineers
   b. Industrial microbiologists
   c. Virologists
   d. Immunologists
   e. Epidemiologists

2. Microbes that are newly discovered human pathogens cause _____.
   a. fungal diseases
   b. emerging diseases
   c. viral diseases
   d. ubiquitous diseases
   e. genetically modified diseases

3. In modern times organisms are classified based on their _____.
   a. structural similarities
   b. similarities in physiology
   c. morphology
   d. location of discovery
   e. similarities in genetics

4. The process of arranging organisms into orderly groups is called ___.
   a. classification
   b. nomenclature
   c. bioremediation
   d. identification
   e. serology

5. Which of the following are procaryotic?
   a. bacteria
   b. viruses
   c. fungus
   d. algae
   e. humans

6. The worldwide death toll of infectious disease is approximately ____ people per year.
   a. 1 million
   b. 80 million
   c. 13 million
   d. 1 billion
   e. 200 million

7. Ancient procaryotes added _____ to an atmosphere that had very little ____.
   a. water : nitrogen
   b. oxygen : oxygen
   c. carbon: water
   d. carbon: oxygen
   e. hydrogen: hydrogen

8. The germ theory of disease states that ____.
   a. microbes can form spores
   b. microbes are procaryotic
   c. antibiotics kill bacteria
   d. diseases may be caused by infection
   e. microbes may infect abiotically

9. Most of the world's photosynthesis is done by _______.
   a. trees
   b. flowering plants
   c. agricultural plants
   d. animals
   e. microorganisms

10. Which of the following is correctly ordered, from smallest to largest?
    a. viruses, fungus, bacteria
    b. bacteria, viruses, fungus
    c. fungus, viruses, bacteria
    d. viruses, bacteria, fungus
    e. bacteria, fungus, viruses

11. Which of the following may be pathogens?
    a. viruses
    b. yeasts
    c. bacteria
    d. all of the above
    e. none of the above

12. Antonie van Leeuwenhoek was a pioneer in the field of ____.
    a. aseptic technique
    b. sterilization
    c. microscopy
    d. spontaneous generation

13. Which of the following are procaryotic domains?
    a. Archaea and Monera
    b. Bacteria and Archaea
    c. Monera and Eukarya
    d. Eukarya and Archaea
    e. Monera and Bacteria

14. A good hypothesis must be supported or discredited by _______
    a. careful thought
    b. repeated inferences
    c. observations or experiments
    d. popular opinion
    e. ancient theories

15. Which of the following is written correctly?
    a. E. COLI
    b. *E. coli*
    c. E. Coli
    d. e. Coli
    e. *e. coli*

Applications of Chapter 1

1. Astrobiologists are scientists that look for signs of extraterrestrial life; rather than looking for "little green monsters" of B-movie fame, they look for signs of procaryotic life. Why do they look for procaryotes and not eukaryotic cells and what are some signs of prokaryotic life you might find in meteorites or rock samples?

Chapter 2 The Chemistry of Biology

Building Your Knowledge

1. What is the smallest piece of an element that still maintains the properties of that element?

2. Draw a helium (He) atom, labeling protons, neutrons, and electrons.

3. How does atomic number (AN) differ from mass number (MN)?

   If we add an electron to an element, do we change its mass number or atomic number? Explain.

   If an uncharged atom has 12 protons, how many electrons does it have?

4. What are elements that have the same atomic number, but different atomic masses called?

5. Which part(s) of an atom are lost, gained, or shared during chemical bonding?

6. If an atom has 6 valence electrons and the valence shell can hold 8 electrons, how many electrons can it accept?

   How many bonds can it participate in?

7. Compare and contrast ionic and covalent bonds.

   In which type of bond are electrons given and taken?

   In which type of bond are the electrons shared?

8. Name one polar compound. Most lipids are nonpolar, what does that mean?

9. How do cations differ from anions? How are cations and anions similar?

10. How does hydrogen bonding differ from ionic bonding?

11. Differentiate between molecular and structural chemical formulas.

Do fructose and glucose have the same molecular formula or the same structural formula?

12. Write a chemical reaction combining molecule A with B to make compound C. Label the products and the reactions. Is this a synthesis, decomposition, or exchange reaction?

13. You make a glass of lemonade from powder. What is your solvent?

What is your solute and what do you call your solution?

14. What are amphipathic molecules?

Are hydrophobic molecules polar or nonpolar?

Which molecules (polar or nonpolar) dissolve easily in water.

15. If you dissolve 5 grams of salt in 100 mL of water, what is the concentration of the solution (in %)?

If you dissolve 5 grams of the same salt in 50 mL of water, is the concentration higher or lower than the first solution?

16. An acid releases ______ when dissolved in water. A base releases ______.

If a solution has a pH of 3, does it have more or less H+ ions than a solution with a pH of 5?

Is the pH of 3 solution more or less acidic than the pH of 5 solution?

17. Why is carbon considered a fundamental element of life?

What types and how many bonds can a single carbon atom participate in?

18. Define and give examples of functional groups.

19. What are the four macromolecules commonly found in living systems?

20. If you discover a new compound and call it newbose, you are telling the world this is a rule.

21. What is dehydration synthesis and how is it related to the building of carbohydrates? How are carbohydrates broken down?

22. Why don't oil and water mix?

What are lipids used for in cells and why is it important that they don't dissolve in water?

23. What subunits combine to form lipids ?

What do lipases do to lipids?

24. Draw a cell membrane, labeling the phospholipids (head and tail), proteins, hydrophobic regions, and hydrophilic regions.

25. Why is it important for membranes to be selectively permeable? What would happen if they were not permeable or were permeable to everything?

26. Draw an amino acid, labeling the amino group, the carboxyl group, and the R group. How are amino acids linked together?

27. Are peptide bonds ionic, covalent, or hydrogen bonds?

28. Describe four separate functions proteins have in cells.

29. What are DNA and RNA made of? List two ways DNA is different from RNA.

30. Draw a nucleotide. Label the nitrogen base, pentose sugar, and the phosphate group.

31. What is the difference between a purine and a pyrimidine? Which bases are purines? Which are pyrimidines?

32. The sugar-phosphate backbone of a DNA molecule is held together by ______________ bonds. The nitrogen base "rungs" of the DNA ladder are held together by ________________________ bonds.

33. What are the three major types of RNA? What is the function of each?

34. How is ATP similar to DNA and RNA? How is it different?

Organizing Your Knowledge

Table 1

Macromolecule	**Subunits**	**Use**	**Examples**
Carbohydrates	a.	Storage	b.
c.	Fatty acids and glycerol	d.	e.
Proteins	f.	g.	Enzymes
h.	Nucleotides	i.	DNA

Table 2

Carbohydrate	Source	Use	Molecular Structure
Agar	Seaweed	a.	Sulfur-conjugated carbohydrates
b.	Fungus and arthropods	Exoskeletons	c.
Peptidoglycan	d.	Cell walls	e.
f.	Plant cell walls	g.	Fibrous long chains of carbohydrates
Starch or glycogen	h.	i.	Long-chain carbohydrates

Table 3

Level of Protein Structure	Description
Quaternary	a.
b.	The order of amino acids in a protein chain
Secondary	c.
d.	Bonds between functional groups (e.g., disulfide bonds)

Table 4

Lipid	Common Function	Structure
Phospholipids	a.	Amphipathic
b.	Reinforces cell membranes	Rings (steroid)
Prostaglandins	c.	Fatty acid derivatives

Self-Test: Vocabulary

ACROSS

2 macromolecule made up of fatty acids and glycerol
5 positively charged particle found in the nucleus of an atom
6 strong bond consisting of electrons shared between two atoms
7 compound that releases H+ in an aqueous solution
13 macromolecules made up of polymers of sugar molecules
14 bonds that hold sugar monomers together
15 sugar found in RNA but not DNA

DOWN

1 the formation of helices or sheets by hydrogen bonding along the chain of amino acids in a protein
3 is water polar or nonpolar?
4 compounds that have both carbon and hydrogen
8 protein that catalyzes chemical reactions
9 bond that holds amino acids together to form a protein
10 outer orbitals of electrons in an atom
11 liquid medium that a solute is dissolved in
12 five carbon sugar

Self-Test: Multiple Choice

1. Hydrophobic molecules are ______ and ________ dissolve easily in water.
   a. nonpolar:do not
   b. polar: do not
   c. nonpolar: do
   d. polar: do
   e. anionic:do

2. If a substance gives off hydroxyl (OH-) ions when dissolved in water, it has a ___ pH and is called a(n) _____.
   a. high: acid
   b. low: acid
   c. high: alcohol
   d. low: base
   e. high: base

3. The charged particles within the nucleus of an atom are called ____.
   a. neutrons
   b. DNA
   c. electrons
   d. glycosides
   e. protons

4. If a bond forms where one atom loses electrons and one gains electrons, the bond is called a(n) ____ bond.
   a. covalent
   b. hydrogen
   c. ionocovalent
   d. ionic
   e. inorganic

5. If two atoms have the same atomic number but different mass numbers, they are described as ____ of the same element.
   a. isomers
   b. orbitals
   c. valences
   d. isotopes
   e. ions

6. Disulfide bonds between cysteine molecules are an example of ___ protein structure.
   a. primary
   b. secondary
   c. tertiary
   d. quanternary
   e. duplicative

7. DNA molecules lack __________, which RNA molecules have.
   a. nitrogenous bases
   b. a 2' OH
   c. phosphate groups
   d. a 3' OH
   e. purines

8. If you discover a new hexose, you have discovered a _____ with six carbons.
   a. protein
   b. lipid
   c. nucleic acid
   d. sugar
   e. amino acid

9. If an atom has five electrons in its valence shell, how many chemical bonds can it participate in?
   a. five
   b. one
   c. three
   d. none, it's full
   e. four, but only if they are double bonds

10. Proteins can be _____.
    a. enzymes
    b. toxins
    c. antibodies
    d. all of above

11. Which of the following is NOT a type of RNA commonly found in procaryotic cells?
    a. messenger RNA
    b. nRNA
    c. ribosomal RNA
    d. mRNA

12. Which of the following elements is NOT found in every amino acid?
    a. nitrogen
    b. sulfur
    c. oxygen
    d. hydrogen
    e. carbon

13. An anion gains its negative charge by _________.
    a. losing electrons
    b. gaining protons
    c. losing protons
    d. gaining electrons

14. Which of the following is in the correct order, from greatest to least strong bonds?
    a. ionic-covalent-hydrogen
    b. covalent-hydrogen-ionic
    c. hydrogen-covalent-ionic
    d. ionic-hydrogen-covalent
    e. covalent-ionic-hydrogen

15. Phospholipids make up a large portion of cell membranes. They are amphipathic, meaning ____.
    a. they have both proteins and lipids attached to them
    b. they have two carbon rings, not one carbon ring
    c. they have both hydrophobic and hydrophilic portions

Applications of Chapter 2

All life on Earth is considered carbon-based. What attributes of carbon make it well suited to be the basis of all biological macromolecules?

Chapter 3 Tools of the Laboratory

Building Your Knowledge

1. Please briefly describe each of the "Five Is" of microbiology.

5 Is	Description
a. Inoculation	
b. Incubation	
c. Isolation	
d. Inspection	
e. Identification	

2. How are streak plate, pour plate, and spread plate techniques similar?

Draw the pattern growth you would expect from each technique.

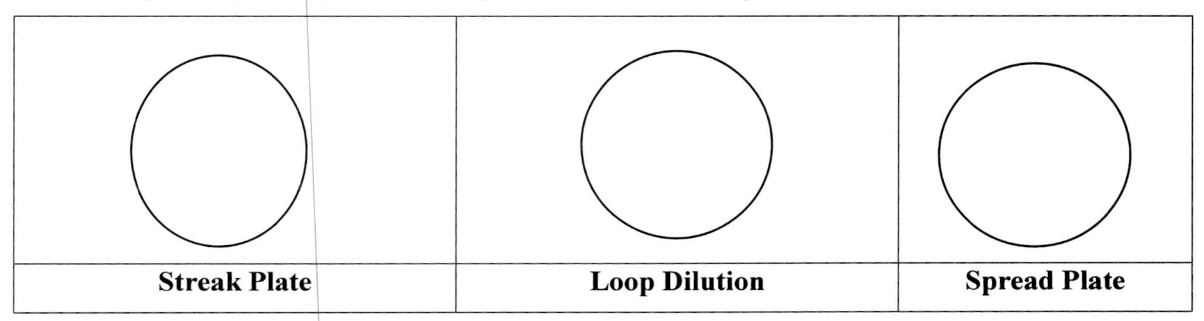

Streak Plate	Loop Dilution	Spread Plate

3. Can you isolate bacteria in a mixed culture more easily using liquid or solid media?

   If you were given a flask with three different bacterial species in it, what is the first thing you would do to separate the three?

4. What is agar and why is it so commonly used in microbiology labs?

5. How do simple media differ from complex media?

   If a media recipe calls for milk or beef extract, are you making a synthetic or complex medium?

6. What is the difference between selective and differential media?

If you use a petri plate with media that changes your target colonies from white to pink, are you using selective or differential media?

7. MacConkey agar is both selective and differential. Explain. It selects for _____________ and differentiates between ____________________ and _______________.

8. What is the difference between a mixed and contaminated culture?

9. Labeling—microscope parts:

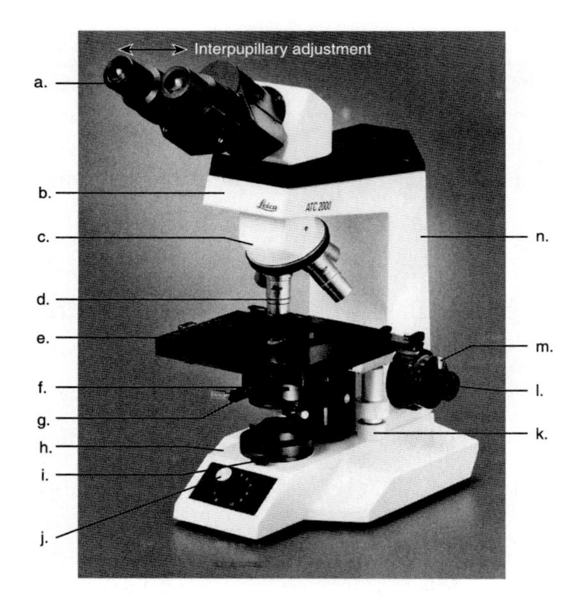

10. How do resolution and magnification differ?

11. If you were a virologist, would you likely be able to see your viral particles with a light microscope? Why or why not?

12. What is confocal scanning optical microscopy, how does it work, and why is it of particular use to scientists studying intracellular bacteria, such as *Listeria monocytogenes*?

13. A hanging drop mount is more work to do than a simple wet mount. When would it be better to do a hanging drop mount?

14. Can you Gram stain living cells and look at the living cells under the microscope? Why or why not?

15. What is the difference between positive and negative staining? Give an example of each type of stain.

16. Describe the process of a Gram stain. In this description include the primary dye and the counterstain used and what positive and negative samples look like under the microscope.

17. What color are gram-negative bacteria when stained? ____________________________

    What color are gram-positive cells? ___________________________________

18. Which bacteria are easily identified with an acid-fast stain? (Hint: These are the acid-fast bacteria.)

19. Name three structures that can be identified by staining.

a. ______________________________

b. ______________________________

c. ______________________________

Organizing Your Knowledge

Table 1

Media Type	Characteristics	Common Use
a.	Clotlike consistency	Motility
Liquid, semisolid, liquefiable solid	Media with complex growth factors or nutrients added.	b.
Reducing media	c.	Used to study growth with or without oxygen
d.	Differential media	Identification of types of hemolysis
Transport media	e.	Used to preserve samples but not allow growth
f.	Contains sugars to be fermented and a pH indicator	Used to identify and characterize bacteria
MacConkey agar	g.	Isolates gram-negative enterics
h.	Media with undefined ingredients, like blood, milk, extracts, infusions.	Growth of fastidious organisms
Mannitol salt agar	Contains 7.5% sodium chloride which inhibits most bacterial growth	i.

Table 2

Type of Microscopy	Illuminating Source	Description	Use
a.	Visible light	Lit specimens surrounded by dark background	b.
Phase-contrast	c.	Changes in density are translated to changes in light intensity	d.
e.	UV radiation and dyes that fluoresce when activated	f.	Labeling structures or bacteria and visualizing them
Scanning electron	g.	Shows surfaces of metal-coated objects in great detail	h.

Table 3

Mount or Stain Technique	Live or Dead Specimens?	Characteristics	Use or Examples
a.	Live	A drop or two of sample with a coverslip	Visualizing the living activities of microbes
Hanging drop	b.	Sample in a drop of water in a depression slide	c.
d.	Dead	Use of a beam of electrons as a light source	Shows fine detail of internal cellular structures and viruses
Capsule stain	Dead	Negative stain with India ink	e.
Differential stains	Dead	f.	Gram stains, acid-fast stains
g.	Dead	Dye is forced (with heating) into stain-resistant structures	To distinguish spores from vegative cells
h.	i.	A dye technique that stains the background, but not the cells	A capsule stain is an example of this type of stain
j.	Dead	Crystal violet primary stain and safranin as counterstain	Commonly used to characterize clinical samples

Self-Test: Vocabulary

Please fill in the blanks below, using your text as a reference.

1. __________ of a medium is the first step in obtaining a pure culture.
2. Isolation of microbes requires that plates have discrete __________.
3. The _______ plate technique uses a "hockey stick" to evenly distribute bacterial cells on a plate.
4. Placing microbes in the right conditions to allow growth of visible colonies is called _____________.
5. A ___________ culture can ruin an analytical test or lead to an incorrect identification of a culture.
6. Media that inhibit the growth of nontarget organisms are _________________ media.
7. If you wanted to move a sample from the patient's bedside to a lab in a different city, you would use _______ media.
8. The ___________ power of a microscope is a measure of the clarity of the image.
9. A capsule stain is an example of a _____________ stain.
10. Media whose chemical composition is unknown, such as milk agar are considered __________ media.

Self-Test: Multiple Choice

1. Media that grows as many microbes as possible is generally called ____.
   a. selective media
   b. differential media
   c. reducing media
   d. general-purpose media
   e. fastidious media

2. A Petri plate with an unknown fungus growing on it is said to be ____
   a. a mixed culture
   b. an axenic culture
   c. a contaminated culture
   d. a subculture
   e. a gnotobiotic culture

3. Which of the following is NOT one of the "Five Is" culturing microbes?
   a. inspection
   b. incubation
   c. isolation
   d. irradiation
   e. inoculation

4. Which of the following staining techniques is NOT a differential staining process?
   a. Gram staining
   b. endospore staining
   c. Loeffler's blue staining
   d. acid-fast staining
   e. all of the above are differential stains

5. If you are making a liquid medium that calls for the addition of meat extract, how could you describe the finished product?
   a. complex agar
   b. synthetic gelatin
   c. complex broth
   d. complex gelatin
   e. synthetic broth

6. The hanging drop slide preparation is commonly used to determine ____.
   a. the species of bacteria you are looking at
   b. the growth rate of the bacteria you are looking at
   c. whether your bacteria are gram-negative or gram-positive
   d. the mobility of the bacteria you are looking at
   e. which antibiotics your bacteria are sensitive to

7. A differential stain uses ____.
   a. a single basic dye to visualize bacterial cells
   b. a single acidic dye to visualize bacterial cells
   c. green capsular dye to visualize viral cells
   d. a primary dye and counterstain
   e. heavy metal dyes to coat target cells for electron microscopy

8. The clarity with which you can see an image is a measure of the ____ of the microscope.
   a. resolution
   b. illumination index
   c. magnification
   d. virtual aperture
   e. field strength

9. If you wanted to isolate several bacterial species from a mixed culture which culture method would you LEAST likely use?
   a. streak plate
   b. spread plate
   c. broth culture
   d. pour plate
   e. nutrient agar

10. Which of the following structures can you NOT commonly see with a light microscope and staining techniques?
    a. bacterial cells
    b. flagella
    c. capsules
    d. bacterial shape
    e. viral particles

11. Acidic dyes are most commonly used to _______
    a. positively stain bacteria for light microscopy
    b. negatively stain bacteria for light microscopy
    c. positively stain fungus for electron microscopy
    d. negatively stain viruses for light microscopy
    e. positively stain viruses for light microscopy

12. A microscope condenser ____.
    a. magnifies an object
    b. focuses light on an object
    c. creates a real image
    d. creates a virtual image
    e. is not part of a compound microscope

13. MacConkey agar is selective because it ________.
    a. can differentiate between *Salmonella* and *E. coli*
    b. encourages enteric bacterial growth while killing other bacteria
    c. has blood added so *Neisseria* will grow
    d. encourages lactobacilli to grow and kills other bacteria
    e. can differentiate between lactobacilli and coccobacilli

14. Semisolid media is useful for____.
    a. determining growth of anaerobes
    b. testing bacterial mobility
    c. determining optimum growth temperatures
    d. determining optimum growth pH
    e. providing micronutrients to growing cultures

15. Which type of microscopy could you NOT use to see living cells?
    a. bright field
    b. dark-field
    c. phase-Contrast
    d. confocal scanning optical microscopy
    e. electron microscopy

Applications of Chapter 3

1. It is estimated that less than 1% of <u>*all bacteria*</u> are culturable. As described in Insight 3.1, there are also many viable nonculturable bacteria in and on the human body. Why was this a surprise to researchers, considering the earlier findings that environmental microbes are rarely culturable?

2. Before agar, gelatin was the common medium used to grow isolated colonies, why is agar better than gelatin?

3. Transport media often stabilizes the cellular structure of microbes, but doesn't allow for the growth of bacterial cells. Why is this important?

Chapter 4 An Introduction to Cells and Procaryotic Cell Structure and Function

Building Your Knowledge

1. All organisms are made up of cells. What are the two fundamental types of cells?

2. What are the seven basic characteristics that living things share?

3. Label the following diagram of a procaryotic cell.

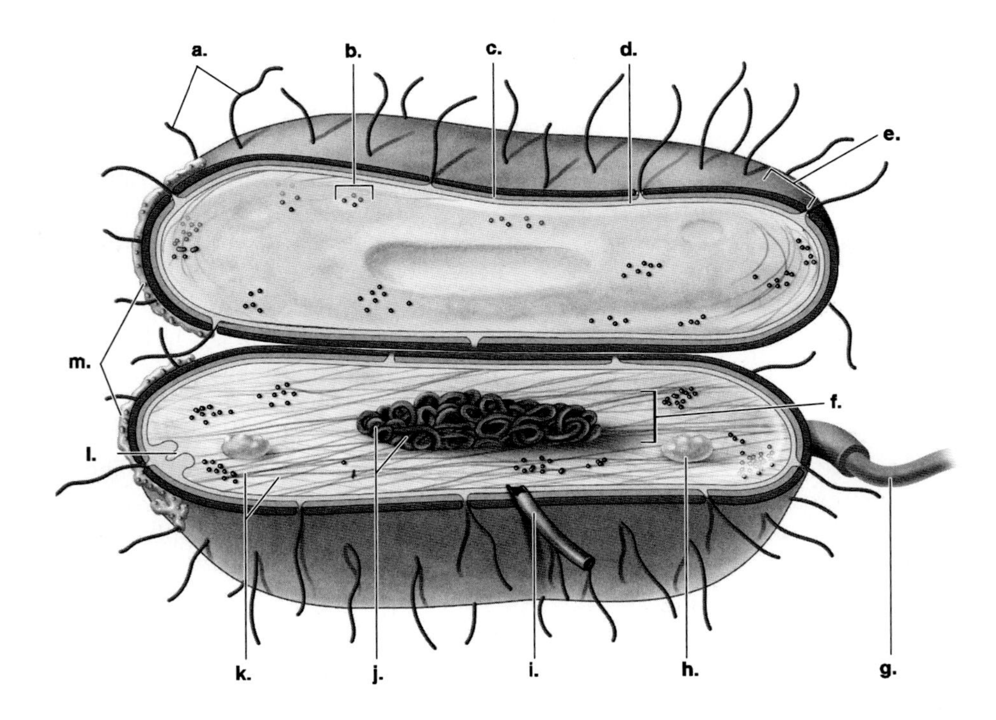

4. Appendages serve two generalized functions for bacteria. What are they?

   a. ______________________________

   b. ______________________________

5. Draw a bacillus with the following flagellar arrangements.

A. Peritrichous	C. Monotrichous
B. Lophotrichous	D. Amphitrichous

6. How do chemical attractants affect the tumble/run cycle of a motile bacterial cell?

7. In what way are the spirochete flagella unusual? How do spirochetes move?

8. What is the difference between pili and fimbriae? Both are used for _____________.
   Pili are used for __________________ .

9. Compare and contrast the structure and functions of a slime layer and capsule.

10. What is peptidoglycan and where is it found in bacterial cells?

    How are the actions of lysozyme and penicillin similar?

11. Which bacteria are generally harder to kill—gram-negative or gram-positive?

12. What is an acid-fast stain used for and which species of bacteria are acid-fast?

13. Why do mycoplasmal cell membranes contain high levels of sterol molecules?

14. List and describe four separate functions of bacterial cell membranes.

    a. ____

    b. ____

    c. ____

    d. ____

15. There are two structures made of DNA in the bacterial cell.

Which is larger and contains essential genes? ____________________

Which genes are commonly found on plasmids?

16. What is the function of ribosomes?

17. What is an endospore?

Why is it an advantage for bacteria to have them?

Name two bacteria that form endospores. __________________ ______________

Are endospores used for reproduction? Explain your answer.

18. List the three most commonly seen bacterial cell shapes.

a. ______________________

b. ______________________

c. ______________________

19. Draw the following arrangements of bacterial cells.

Diplococci	Streptococci
Staphylococci	Tetrad

20. Give two reasons why microbiologists are concerned with classifying bacterial groups.

21. Why are descriptive tests such as a Gram stain and cell shape not detailed enough to classify microbes?

22. What gene sequences are commonly used to compare and classify newly described bacteria?

23. Phenetic classification schemes group bacteria based on ___________________.
Phylogenetic classification schemes group bacteria based on _________________.

24. The current edition of *Bergey's Manual of Determinative Bacteriology* separates bacterial groups based on _________________.

25. Which system, phenotype or phylogeny, do medical microbiologists commonly use to identify bacteria in clinical samples? Why?

26. What is a strain of bacteria?

27. How are cyanobacteria and green and purple sulfur bacteria similar?

Which group produces oxygen in the process?

28. What are myxobacteria and why are they different from most bacteria?

29. Name two bacterial obligate intracellular parasites and the diseases they cause.

a. ___________________________

b. ___________________________

Can obligate intracellular parasites be grown on general media agar plates? Why or why not?

30. Which procaryotic domain is most closely related to eucaryotes?

Give several examples of extreme environments where Archaea have been found.

Organizing Your Knowledge

Table 1

Characteristic of Living Things	**Means to Achieve This Characteristic**
1. _____ movement	a. binary fission
	b. cell walls
2. _____ reproduction	c. chromosomes or plasmids
	d. cilia or flagella
3. _____ heredity	e. enzymes
	f. Golgi apparatus
4. _____ metabolism	g. granules or inclusions
	h. mitochondria
5. _____ cell support and protection	i. ribosomes
	j. semipermeable membranes
6. _____ transport	k. sexual or asexual processes
	l. synthesis reactions

Table 2. Fill out the following table, describing the structures found in procaryotes and the function of each.

Structure	**Location**	**Function**
Capsule	External	a.
Flagella	b.	c.
d.	Entire cell	Environmentally resistant form
Actin cytoskeleton	e.	f.
g.	External	Adhesion
Inclusions	h.	i.
j.	Internal (cytosol)	Protein synthesis (translation)
k.	l.	Protection from hypotonic lysis

Self Test: Vocabulary

ACROSS

4 small circular piece of DNA that carries non-essential genes
7 domain of prokaryotes that has many extremophiles
9 storage structure
10 organisms that lack membrane-bound organelles and a nuclear membrane
11 color of stained gram-positive cells
12 extermal structure - either the slime layer or capsule
13 appendages used for motility
14 form of bacteria that is resistant to environmental stress

DOWN

1 appendages required for conjugation
2 sum of all the chemical reactions in a cell
3 obligate intracellular parasite that causes Rocky Mountain Spotted Fever
5 group of bacteria that lack a cell wall and have sterols in their plasma membrane
6 polymer of sugars linked together with short peptide chains
7 filaments that are found in the cytosol and maintain cell shape
8 periplasmic flagella are found in ___.

Self-Test: Multiple Choice

1. A capsule is used by bacterial cells for all of the following EXCEPT:
    a. conjugation
    b. protection against phagocytes
    c. adhering to surfaces
    d. formation of biofilms

2. Which of the following statements is FALSE, concerning bacterial cell walls?
    a. they have peptidoglycan
    b. they give cells their shape
    c. they protect the cell from hypertonic lysis
    d. they are the target of penicillin action

3. A flagellum is used by a bacterial cell for:
    a. adhesion
    b. structural support
    c. protein synthesis
    d. motility

4. Archaeabacteria include ___.
    a. many human pathogens
    b. mostly flagellated bacteria
    c. extremophiles
    d. all the gram-negative bacterial species

5. A flagellated bacterial cell moving toward a food source will ___.
    a. make a straight line right for the food
    b. tumble more than it runs
    c. run more than tumble
    d. shed its flagella and move with its slime layer

6. If you Gramstain a culture and see purple circles arranged in chains, you would call them:
    a. gram-negative bacilli
    b. gram-positive staphylococci
    c. gram-negative staphylococci
    d. gram-positive streptococci

7. Bacteria are taxonomically classified by___.
    a. cell shape
    b. rRNA sequence similarity
    c. mechanism of mobility
    d. colony morphology

8. Bacterial plasmids will likely carry all the following genes EXCEPT:
    a. the gene to use a different sugar source
    b. antibiotic resistance genes
    c. genes for the proteins required in metabolism
    d. all of these are commonly seen on plasmids

9. Which of the following structures is NOT found in the cell envelope of a bacterial cell?
    a. cell wall
    b. ribosomes
    c. capsule
    d. glycocalyx

10. If a bacterial cell lost its ribosomes, it would no longer be able to ___.
    a. produce proteins
    b. produce DNA
    c. produce lipids
    d. produce a flagella

11. Which of the following bacteria are photosynthetic?
    a. Cyanobacteria
    b. Chlamydia
    c. Pseudomonas
    d. Treponema

12. Gram-positive cell walls ___.
    a. contain LPS
    b. have a thick layer of peptidoglycan
    c. have porins
    d. have an outer membrane

13. Smooth, encapsulated bacteria are generally less pathogenic than are rough bacterial strains.
    a. True
    b. False

14. Which group of bacteria have periplasmic flagella?
    a. bacilli
    b. cocci
    c. vibrio
    d. spirochetes

15. Endospores are used by some bacterial species to reproduce.
    a. True
    b. False

Applications of Chapter 4

The formation of biofilms on medical devices is of concern to physicians and patients alike. What are some traits of biofilms that make them difficult to treat?

Chapter 5 Eucaryotic Cells and Microorganisms

Building Your Knowledge

1. What is intracellular symbiosis and how does in relate to the evolution of a eucaryotic cell?

2. Describe the movement from simple independent, single-celled eucaryotes to complex multicellular eucaryotes. Which came first, specialization of cell functions or formation of colonies?

3. Label the structures found in a "composite" eucaryotic cell. These structures aren't ALL found in any specific eucaryote, but most are found in most eucaryotic cells.

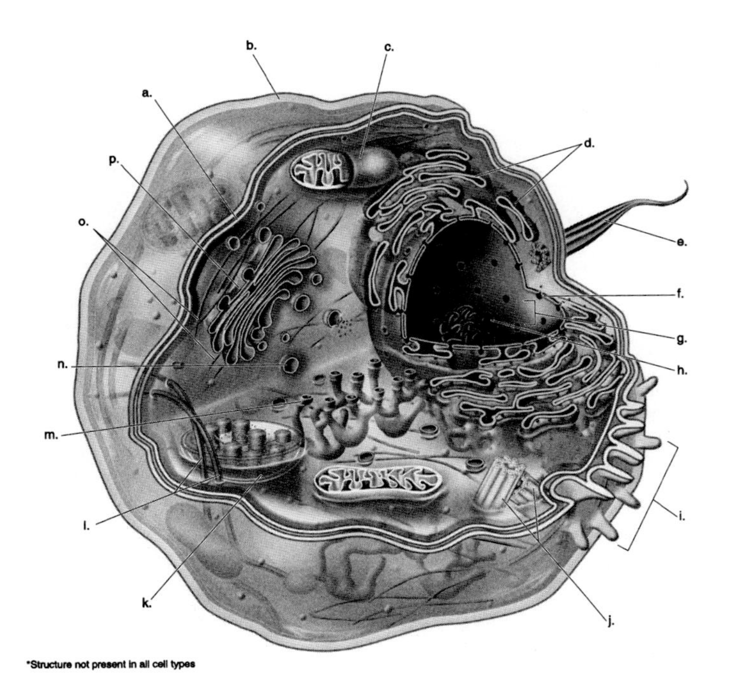

4. Why is mobility important for some organisms?

   What two structures may propel a protozoan?

5. Are procaryotic and eucaryotic flagella identical in structure and function? Explain your answer.

6. Where is the eukaryotic glycocalyx and what is the function of a glycocalyx?

7. Fungal cell walls are made up of ____________________.

   Do algae and fungi have the same types of cell walls? Explain.

8. One distinguishing characteristic of eucaryotic cells is the presence of a nucleus. Draw a nucleus, labeling the nuclear pores, chromosomes, and nucleolus.

9. What is the difference between rough endoplasmic reticulum and smooth endoplasmic reticulum?

10. Trace the path of a protein for export from the nucleus (where mRNA is formed. to the outside of the cell, labeling the major structures and processes the protein would pass.

11. If a cell lacked lysosomes, what would it not be able to do?

    Do you think amoebas and other phagocytes have more or fewer lysosomes than would photosynthetic algae? Explain.

12. Draw a mitochondrion, labeling its inner membrane, outer membrane, matrix, and cristae.

13. How are mitochondria and chloroplasts similar?

14. What two cytoskeletal elements are found in eucaryotic cells?

Which moves cilia and flagella?

Which moves pseudopods?

15. How have fungi been classified in the past?

How are they classified now?

16. Differentiate between yeasts, hyphae, and pseudohyphae.

What are dimorphic fungi?

17. How do fungi acquire nutrition?

Are they autotrophic or heterotrophic?

18. Are most fungi fragile or relatively hardy growers?

19. If someone or something has a mycosis, what are they infected with?

20. The body of a fungus is called a ___________.

21. What is the difference between:

Septate and nonseptate hyphae?

Vegetative and reproductive hyphae?

22. Do bacterial endospores and fungal spores serve the same purpose?

23. What is the difference between sporangiospores and condida?

24. Are zygospores produced sexually or asexually?

What is the advantage of sexual reproduction?

25. Different fungi have different structures for sexual reproduction. How do zygospores, ascospores, and basidospores differ?

26. The Mastigomycota live in the __________ and cause disease in _____________

and ____________________.

27. The Amastigomycota are split into four divisions based on _________________.

28. What is different about the Deuteromycota, compared to the other three divisions of Amastigomycota?

29. List three types of media that are generally useful in growing fungus.

a. ____________________________

b. ____________________________

c. ____________________________

30. Is spore formation generally used in clinical laboratory identification of fungal cultures?

31. Fungal infections are a particular problem for immunocompromised patients. Why?

32. Why are fungi of concern to food microbiologists?

33. What is the difference between algae and protozoans?

34. Where have algae been found growing?

35. What are the five divisions of algae and their common names?

   a. ____

   b. ____

   c. ____

   d. ____

   e. ____

36. Why are dinoflagellates of medical importance to humans?

37. How do protozoans get their nutrients? Are they photosynthetic or heterotrophic?

38. How do protozoans move? What structures may be used?

39. What is the difference between trophozoites and cysts? If a pathogen does not form a cyst, would you expect it to be easily transmitted in food or water? Why or why not?

40. Name the four groups of parasitic protozoans and give an example of each.

41. How are most clinically relevant human parasites identified in clinical labs?

42. How are trypanosomes spread?

    What diseases do they cause?

43. How is amebic dysentery spread?

What are the major symptoms of the disease?

44. What are helminths?

Are the medically important helminths free-living or parasitic?

Which structures are most pronounced in helminths (circulatory, nervous, reproductive, digestive, etc.)?

45. How are helminth infections identified in the laboratory?

Organizing Your Knowledge

Table 1

Fungal Group	Sexual Reproductive Structures	Common Examples
Zygomycota	a.	e.
Ascomycota	b.	f.
Basidiomycota	c.	g.
Deuteromycota	d.	h.

Table 2

Structure	Found in: Eucaryotes (E), Procaryotes (P), or Both(B)	Function?
Smooth endoplasmic reticulum	a.	1.
Flagella	b.	2.
Nucleus	c.	3.
Rough endoplasmic reticulum	d.	4.
Glycocalyx	e.	5.
Golgi apparatus	f.	6.
Cilia	g.	7.
Mitochondria	h.	8.
Cell wall	i.	9.
Plasma membrane	j.	10.
Nucleolus	k.	11.
Chloroplasts	l.	12.
Ribosomes	m.	13.

Self-Test: Multiple Choice

1. Red tides are caused by overgrowths of ___.
   a. red algae
   b. diatoms
   c. dinoflagellates
   d. apicomplexans

2. A medical mycologist studies ________.
   a. mycorrhizae on plant roots
   b. fungus than infect humans
   c. algae that infect fish
   d. yeasts that are used to brew beer

3. Most water-transmitted protozoan parasites have a well-developed ____ stage.
   a. trophozoite
   b. cyst
   c. endospore
   d. sarcodinan

4. Amebic dysentery is generally transmitted by ____.
   a. self-inoculation
   b. eating moldy bread
   c. eating contaminated food or water
   d. ticks or fleas

5. The cell walls of fungus are made up of _____________.
   a. chitin
   b. cellulose
   c. silicon dioxide
   d. cellulose
6. Which eucaryotic organelles were likely originally free-living bacteria?
   a. chloroplasts and mitochondria
   b. nucleus and lysosomes
   c. Golgi apparatus and nucleus
   d. nucleolus and flagella
7. Protozoa may move by all of the following EXCEPT_____.
   a. pseudopods
   b. flagella
   c. pili
   d. cilia
8. Conidia differ from zygospores in that ____.
   a. conidia are produced sexually, zygospores are asexually produced
   b. conidia are reproductive structures in bacteria, zygospores are found in fungi
   c. conidia have chitonous cell walls. Zygospores do not
   d. conidia are produced asexually, zygospores are sexually produced
9. Trypanosomes are important protozoan parasites transmitted by ____ and cause ____ .
   a. mosquito: malaria
   b. water: amebic dysentery
   c. insect bite: sleeping sickness
   d. feces: tapeworm infection
10. The nucleolus is the ___.
    a. area where lysosomes are made
    b. site of ribosome synthesis
    c. central part of a procaryotic cell
    d. site of flagellar attachment
11. The Amastigomycota, or terrestrial fungi, are organized based on their ____.
    a. size and color
    b. asexual reproductive mode
    c. digestive mode
    d. sexual reproductive mechanism
12. Which of the following pairs is mismatched?
    a. flagella—movement
    b. lysosomes—protein degradation
    c. cilia—adhesion
    d. ribosomes—protein synthesis
13. Which of the following are NOT eucaryotic?
    a. fungus
    b. algae
    c. protozoa
    d. bacteria
14. Which of the following organelles does a protein NOT go through on its way out of a cell?
    a. lysosomes
    b. rough ER
    c. vesicles
    d. Golgi apparatus
15. Most parasites dedicate a great deal of body space to _______ structures.
    a. reproductive
    b. digestive
    c. circulatory
    d. nervous system

Applications of Chapter 5

1. Antifungal agents, such as Amphotericin B, often have much more severe side effects than antibacterial agents. Based on the knowledge of procaryote and eucaryote cell structure and function, why do you think this is the case?

2. Most known bacteria have cell walls and lack sterols in their membranes. Sterols are commonly found in eukaryotic cell membranes. Which group of bacteria lack cell walls? What supplements do these bacteria require when in culture? What is the function of sterols?

Chapter 6 An Introduction to Viruses

Building Your Knowledge

1. Can you see viral particles with a light microscope?

2. What does the term virus mean in Latin? Who coined the term virus?

3. How were viruses discovered as plant and animal pathogens?

4. What cell types can be infected with viruses?

5. Are viruses alive? Explain your answer.

   Why are viruses called "particles" and not cells?

6. All viruses are obligate intracellular parasites. What does this mean?

   Can viruses be grown on nutrient agar? Why or Why not?

   Compare and contrast *Rickettsia* and a virus.

7. Draw two viral particles, labeling the capsid, nucleic acid, envelope, and spikes.

Naked	Enveloped

8. A viral particle is made up of __________ and ___________.

9. What are capsomers and where are they found on a viral particle?

10. Do all viruses have an envelope?

    What is a viral envelope made up of (DNA, sugar, lipids, protein)?

    Where does the viral envelope come from?

11. What are viral spikes used for?

12. What three functions do capsids and envelopes perform for a viral particle?

    a. ______________________________

    b. ______________________________

    c. ______________________________

13. Are DNA and RNA ever found in the same viral particle?

14. What else may be found inside a viral particle, with capsid proteins, a nucleic acid core, and an envelope?

15. What are some of the major criteria used to classify viruses?

16. How does a viral particle enter a human host cell and how does this determine host range?

17. Where are most DNA viruses assembled in animal cells?

    Where are most RNA viruses assembled in animal cells?

18. Do enveloped viruses lyse animal cells as they exit?

Do naked viruses lyse animal cells?

19. Name two persistent viral infections and the diseases they cause.

a. ____

b. ____

20. How can viral infection lead to cancer formation? Describe two different viruses that have been shown to cause cancer.

21. Differentiate between a lytic and lysogenic viral cycle.

22. Label the figure below with these terms: adsorption, penetration, assembly, maturation, lysis of the cell, and duplication of phage components.

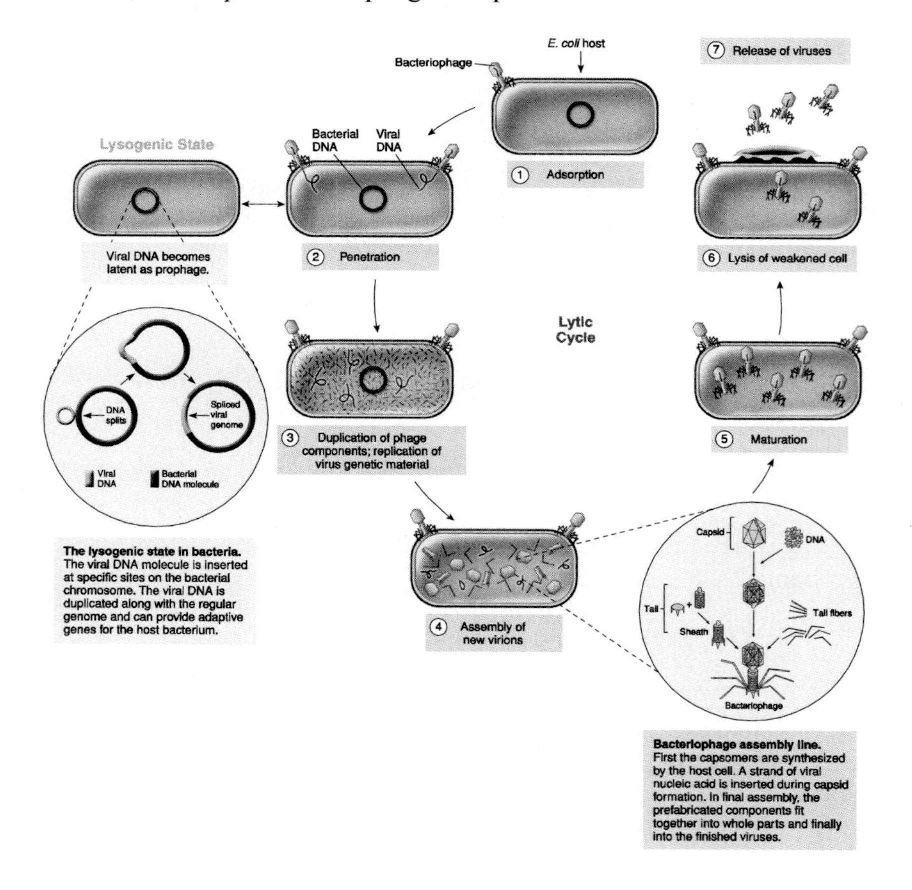

The lysogenic state in bacteria. The viral DNA molecule is inserted at specific sites on the bacterial chromosome. The viral DNA is duplicated along with the regular genome and can provide adaptive genes for the host bacterium.

Bacteriophage assembly line. First the capsomers are synthesized by the host cell. A strand of viral nucleic acid is inserted during capsid formation. In final assembly, the prefabricated components fit together into whole parts and finally into the finished viruses.

23. Define and give an example of lysogenic conversion.

24. Why do scientists cultivate viruses? List three reasons.

25. How are viruses cultured? List three separate ways.

a. ______________________________

b. ______________________________

c. ______________________________

26. List five different diseases caused by viruses.

a. ______________________________

b. ______________________________

c. ______________________________

d. ______________________________

e. ______________________________

27. What are prions?

How do they differ from viruses?

28. What are viriods and how do they differ from viruses and prions?

29. Why are antibiotics ineffective against viral infections?

30. Most antiviral drugs are toxic to the host as well as the virus. Why do you think this is the case?

Organizing Your Knowledge

Structure	**Function**	**Location on Virion**	**Found in ALL Viruses?**
Capsid	a.	b.	c.
d.	Attachment	e.	No—enveloped only
f.	g.	Lipid bilayer coat	h.
i.	Carries genetic information	j.	k.
Reverse transcriptase	l.	m.	n.

Self-Test: Vocabulary

ACROSS

3 binding of virus to host cell
5 protein shell that surrounds viral nucleic acid
6 flexible membrane of host origin found on certain viruses
9 proteins found on enveloped viruses that bind to receptors on host cells
11 release of enveloped virus from animal host cell
15 proteinaceous infectious particle
17 effects of viral infection that cause damage to the host cells
18 enzyme that can synthesize DNA from an RNA template
19 common method of culturing viruses is to inoculate ___.
20 envelope and capsid is removed from a virus, freeing the nucleic acid into the cytosol.

DOWN

1 virulent phage
2 virus that infects bacterial cells
4 virus that is capable of infecting cells of all mammalian species
7 plant infectious agent made up of RNA alone
8 macroscopic "holes" in cell layer caused by viral destruction of host cells
10 inactive viral state seen in lysogenic viral cycles
12 phase of a bacteriophage's cycle that does NOT lyse the cell, but instead forms a prophage
13 activation of lysogenic prophage to begin lytic cycle
14 an obligate intracellular parasite made up of proteins and nucleic acids
16 site of assembly for DNA viruses that infect animal cells

Self-Test: Multiple Choice

1. Where does an enveloped virus's envelope originate?

    a. the host cell's plasma membrane
    b. the capsid of a second virus
    c. the nuclear envelope of bacterial cells
    d. the inner membrane of host cell mitochondria

2. Viruses are commonly visualized using a light microscope on its highest setting.

    a. True
    b. False

3. Transduction is a form of gene transfer in bacteria.

    a. True
    b. False

4. Spikes are found on _________ viruses and are used for _______.

    a. naked viruses: movement
    b. enveloped viruses : adhesion
    c. RNA viruses: adhesion
    d. enveloped viruses: movement

5. If you lyse cells during prophage stage, _____.

    a. many active virus particles will be released
    b. enveloped viruses will become infectious naked viruses
    c. naked viruses will become dormant enveloped viruses
    d. no infective virus particles will be released

6. Which of the following is NOT used to cultivate viruses?

    a. tissue culture cells
    b. eggs
    c. blood agar
    d. mice

7. Lysogenic phage ___.

    a. lyse the host cell immediately after entry
    b. infect cells and integrate their DNA into the host cell's DNA
    c. do not exist in nature
    d. all of the above are correct

8. Which of the following can form a latent viral infection?

    a. herpes virus
    b. common cold
    c. lytic bacteriophage
    d. there are no latent infections in nature

9. A viral particle is made up of ___.

    a. DNA and RNA in the same particle
    b. DNA and lipids
    c. DNA or RNA and proteins
    d. sugars and lipids

10. Tissue tropism determines _______.

    a. which tissues a virus will infect
    b. how long a virus will take to kill a cell
    c. whether a particle is a DNA or RNA virus
    d. which tissues will die first

11. Hepatitis B infects only human liver cells because ____.

    a. only liver cells have enough DNA to infect
    b. only human cells have plasma membranes
    c. only liver cells have the right receptors on their surface
    d. only human cells have a nucleus

12. Where do RNA viruses that infect animal cells mature?

    a. in the nucleus
    b. under the cell wall
    c. in the cytosol
    d. in the chloroplasts

13. Which of the following organisms can serve as a host for a virus?

    a. bacteria
    b. fungus
    c. animal
    d. all of the above

14. Prions are ____.

    a. infectious DNA viruses
    b. dormant RNA viruses
    c. infectious proteins
    d. infectious RNA

15. Enveloped viruses leave animal cells ____.

    a. by disrupting the cell
    b. by budding off the plasma membrane
    c. by disrupting the nucleus which kills the cell
    d. enveloped viruses never leave animal cells

Applications of Chapter 6

1. Most physicians will not send out clinical samples from acute respiratory infections for culture, but may assume the problem is a viral infection based on the clinical profile (symptoms, timing of infection, complete blood count). Why do you think this is the case?

2. How can bacteriophage be used to control bacteria?

Chapter 7 Elements of Microbial Nutrition, Ecology, and Growth

Building Your Knowledge

1. How do essential nutrients differ from nonessential nutrients?

2. Fill out the table below, summarizing the differences between macronutrients and micronutrients.

	Quantities Needed by Cell	**Use in Bacterial Cell**	**Examples**
Macronutrient	a.	c.	e.
Micronutrient	b.	d.	f.

3. How do autotrophs differ from heterotrophs?

   Autotrophs get carbon from ____________________.

   Heterotrophs get carbon from ____________________.

4. How do photoautotrophs and chemoautotrophs differ in how they get energy?

   Photoautotrophs get energy __________________.

   Chemoautotrophs get energy __________________.

5. Cyanobacteria and photosynthetic algae get carbon and energy from __________________.

6. What is the difference between saprobes and parasites/pathogens?

7. How do saprobes acquire nutrients? Where is the organic material digested—on the inside or outside of the cell that is feeding? Explain.

8. Under what conditions have saprobes been found as disease-causing organisms?

9. Can obligate parasites be cultured using solid, synthetic media? Explain your answer.

10. Give two different examples of obligate intracellular parasites.

11. Without the addition of energy, do molecules move from high concentrations to low concentrations or from low concentrations to high concentrations?

12. Draw a cell in hypertonic, isotonic, and hypotonic solutions. Use x's to indicate solute molecules. Which cells shrink? Which swell?

Hypertonic	**Isotonic**	**Hypotonic**

What does a bacterial cell wall protect against?

13. How are facilitated diffusion and active transport similar?

How are they different?

Do both require energy?

14. When is active transport necessary?

What advantage does group translocation have over simple pumps?

15. What are the three cardinal temperatures of microbial growth?

16. What are psychrophiles, mesophiles, and thermophiles?

Which are of concern to food microbiologists?

Which are most commonly pathogenic?

17. How do the oxygen requirements of obligate aerobes, facultative anaerobes, microaerophiles, and obligate anaerobes differ?

18. How would each grow in thioglycolate broth? Draw each test tube.

Obligate Aerobe	Microaerophile	Obligate Anaerobe	Facultative Anaerobe

19. Halophiles live in extreme ____________________ conditions.

20. Barophiles live in extreme ____________________ conditions.

21. How is satellitism a form of commensualism between bacteria?

22. Diagram a way that a nitrogen fixer and cellulose digester can live in synergy?

23. Define and give an example of antagonism in microbial ecology.

24. What are microbial biofilms and how do the different cells of a biofilm "talk" to one another?

25. Why are biofilms of concern to medical microbiologists?

26. How do bacteria reproduce? Draw the process of binary fission below.

27. Do all bacteria reproduce at the same rate? Explain and give examples.

28. Draw a growth curve, labeling lag phase, exponential phase, stationary phase, and death phase.

Time

29. Which stage has the fastest-growing bacteria?

30. Why do cultures move from log phase to stationary phase?

31. If you place 25 bacterial cells in media and their doubling time is 30 minutes, how many cells are in the media at the end of 5 hours?

32. How may we count bacteria (list three ways)?

a. ________________________________

b. ________________________________

c. ________________________________

Live cells only— ________________________________

Organizing Your Knowledge

Table 1

Macronutrient	**Major Use in Bacterial Cells**	**Source or Environmental Reservoir**
Carbon	Carbon-based macromolecules (proteins, lipids, carbohydrates, nucleic acids)	Autotrophs a.
		Heterotrophs b.
Nitrogen	c.	d.
Oxygen	e.	Atmospheric and organic sources
Phosphorus	f.	Phosphate salts
Sulfur	Methionine, cysteine, biotin	g.

Table 2

Type of Bacteria	**Living Conditions Preferred**
psychrophile	a.
b.	Acidic pH
Obligate anaerobe	c.
d.	Small amounts of oxygen
Alkalinophile	e.
f.	Extreme salt conditions
Mesophile	g.
h.	Extreme heat
Obligate aerobe	i.
j.	Extreme pressure (atmospheric)

Table 3. Microbial Symbioses

Relationship	Interaction (+/+, +/–, +/0, etc.)
Mutualism	a.
Commensalism	b.
Parasitism	c.
Synergism	d.
Antagonism	e.

Table 4. The Growth Curve

Stage of Growth Curve	What's Happening?	Growth Speed (Fast/Slow/Level)	Live/Dead Cell Ratio
Lag phase	a.	b.	Most cells live
Exponential phase	Cells reproducing at maximum rate	c.	Most cells live
Stationary phase	d.	Level	e.
Death phase	Toxins building up	Slow	f.

Self-Test: Vocabulary

1. ______the process by which organisms acquire nutrients from the environment
2. ______organism that gets carbon from inorganic sources and energy from the sun
3. ______nutrients that serve as cofactors for enzymes and stabilize protein structure
4. ______organisms that acquire most of their nutrition by decomposing dead material
5. ______chemoautotrophs that produce methane from hydrogen gas and carbon dioxide
6. ______most prevalent macromolecule in an *E. coli* cell.
7. ______an organism that can live with or without oxygen is a _____ anaerobe
8. ______environment where the solute concentration is higher in the environment than it is in a cell
9. ______macronutrient found in proteins and nucleic acids—usable form is ammonia
10. ______organisms that thrive at cold temperatures
11. ______organisms that thrive at low oxygen levels
12. ______organisms that thrive at high pH
13. ______ growth of two organisms on an agar plate where one organism provides nutrition or protection needed by a second factor
14. ______"city of microbes" complex association of mixed microbes with an extracellular matrix
15. ______counting device that counts bacteria by passing them by an electronic detector

A. acidophiles
B. aerobes
C. alkalinophiles
D. biofilm
E. capnophiles
F. carbon
G. chemoautotroph
H. Coulter counter
I. facultative
J. hypertonic
K. hypotonic
L. macronutrients
M. malaria
N. methanogens
O. microaerophile
P. micronutrients
Q. nitrogen
R. nucleic acids
S. nutrition
T. obligate
U. photoautotroph
V. proteins
W. psychrophiles
X. Rickettsia
Y. saprobes
Z. satellitism

Self-Test:Multiple Choice

1. If placed in a hypertonic solution, most bacterial cells will____.

    a. burst if they lack a cell wall
    b. remain unchanged
    c. shrink and die
    d. change color

2. Bacteria preferring low temperatures for optimum growth are called_____.

    a. barophiles
    b. halophiles
    c. thermophiles
    d. psychrophiles

3. Macronutrients are required by cells in ___ quantities and are used to ___.

    a. small: boost enzyme function
    b. large: boost enzyme function
    c. small: form cell structures
    d. large: form cell structures

4. Which of the following methods measures live bacterial cells only?

    a. turbidity
    b. plate counts
    c. cytometer
    d. Coulter counter

5. Which of the following transport processes requires energy?

    a. diffusion
    b. osmosis
    c. facilitated diffusion
    d. group translocation

6. Which of the following microbial associations is NOT a positive symbiosis for both organisms?

    a. mutualism
    b. commensalism
    c. synergism
    d. parasitism

7. Which phase of the growth curve sees an equal rate of bacterial death and reproduction?

    a. lag phase
    b. stationary phase
    c. exponential phase
    d. death phase

8. Strict halophiles are commonly human pathogens.

    a. True
    b. False

9. Photosynthetic bacteria are considered ___.

    a. nonexistent—bacteria don't have chloroplasts
    b. heterotrophs because they feed off dead things
    c. autotrophs because they get their carbon from carbon dioxide
    d. saprobes because they feed off dead things

10. As a bacterial culture grows, the media ______.

    a. gets thicker because of all bacteria
    b. gets cloudier because of all the bacteria
    c. gets warmer because of the heat generated by bacterial cells
    d. gets clearer because the bacteria consume all the nutrients

11. What is the correct order for a growth curve progression, with bacterial cells in batch culture?

    a. lag phase—exponential—stationary
    b. stationary—lag phase—exponential
    c. exponential—stationary—lag phase
    d. lag phase—stationary—exponential

12. Bacteria lacking superoxide dismutase and catalase are ___.

    a. strict aerobes
    b. strict anaerobes
    c. facultative anaerobes
    d. strict acidophiles

13. What are the three cardinal temperatures for microbial growth?

    a. hypertonic, isotonic, and hypotonic
    b. minimum, maximum, and optimum
    c. aerobic, anaerobic, and microaerobic
    d. halophile, barophile, osmophile

14. Active transport mechanisms are required to ___
    a. move nutrients from high concentrations to low
    b. move any nutrient across a plasma membrane
    c. complete facilitated diffusion
    d. move molecules from low concentrations to high

15. Most human pathogens are ______.
    a. mesophiles
    b. psychrophiles
    c. thermophiles
    d. psychrotrophs

Applications of Chapter 7

1. Why are obligate psychrophiles not of concern to food microbiologists, but facultative psychrophiles are?

2. If you were a medical microbiologist studying ways to defeat biofilms, what targets can you think of that would either inhibit biofilm formation or break apart established biofilms?

Chapter 8 Microbial Metabolism

Building Your Knowledge

1. Compare and contrast the two branches of metabolism.

Branch	Processes Occurring	Requires Energy?
Anabolism	a.	c.
Catabolism	b.	d.

2. Fill in the following figure, using figure 8.1 as a reference. Indicate what macromolecules are built, what the sources of energy are, where catabolism occurs, where anabolism occurs, and what energy currency the cell uses.

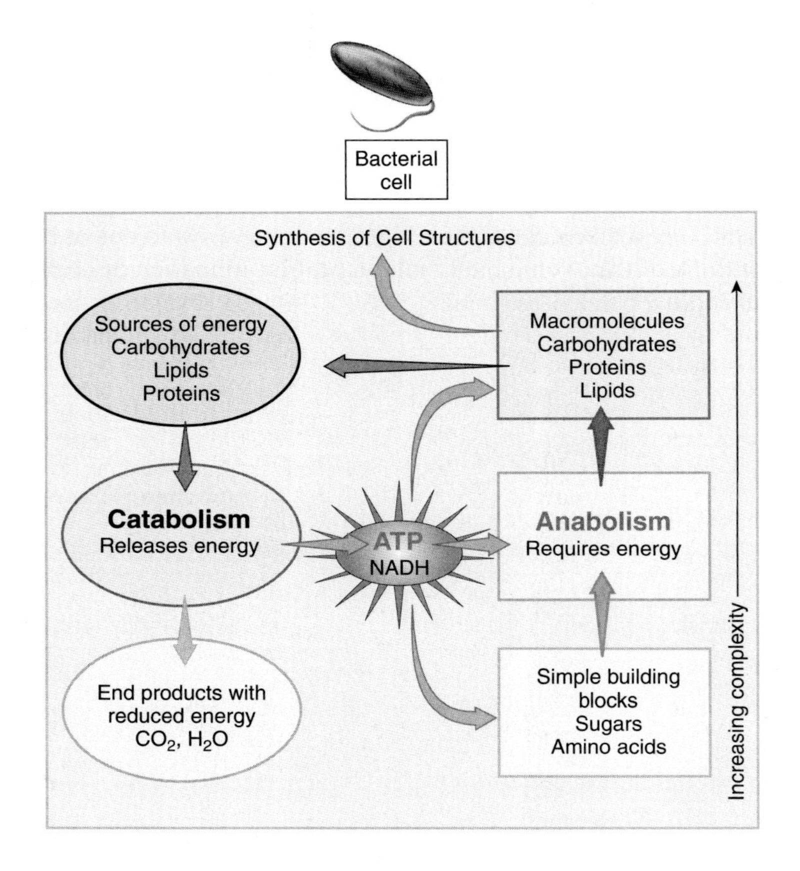

3. What do catalysts do in a chemical reaction?

Do enzymes add energy to chemical reactions?

Are they changed by the reaction?

4. Do they interact with several substrate molecules or one molecule per enzyme (then the enzyme goes away)?

5. What are most biologically relevant enzymes made up of—proteins, lipids, or sugars? ____________________

6. How do enzymes speed chemical reactions?

7. How do holoenzymes differ from apoenzymes?

8. How is the "lock and key" model a good analogy for the interactions of an enzyme with its substrates?

   How is this not an adequate analogy?

9. Endoenzymes work inside the cell. What are enzymes that work outside a cell called?__________________________________

10. Enzymes that are present all the time are called ____________________.

    Induced enzymes are activated or produced only when __________________ is present.

11. The removal of water is a __________________ reaction.

12. The addition of water _______________ chemical bonds.

13. How are oxidation and reduction related?

    If a molecule is reduced, does it gain or lose electrons? _______________

    If a molecule is oxidized, does it gain or lose electrons? _______________

14. How do enzymes contribute to the disease process caused by *Streptococcus pyogenes, Pseudomonas aeruginosa,* and *Clostriduium perfringes?*

Organism	**Enzyme**	**Disease**
Streptococcus pyogenes	a.	d.
Pseudomonas aeruginosa	b.	e.
Clostriduium perfringes	c.	f.

15. How do enzymes lose activity?

16. Are metabolic reactions generally a single step or single reaction? Explain.

17. Describe three separate ways in which enzyme-catalyzed reactions may be regulated.

18. What is energy?

List three forms of energy.

a. ______________________________

b. ______________________________

c. ______________________________

Which forms of energy are most commonly used in cells?

19. How do endergonic and exergonic reactions differ?

Which are typically anabolic? ________________________

Which are typically catabolic?_________________________

20. Which molecule is in reduced form—glucose or carbon dioxide? Why is this significant?

21. What is ATP and why is it called "metabolic money"?

Label the following figure, indicating the location of the three phosphate groups, ribose and nitrogen base (adenine)

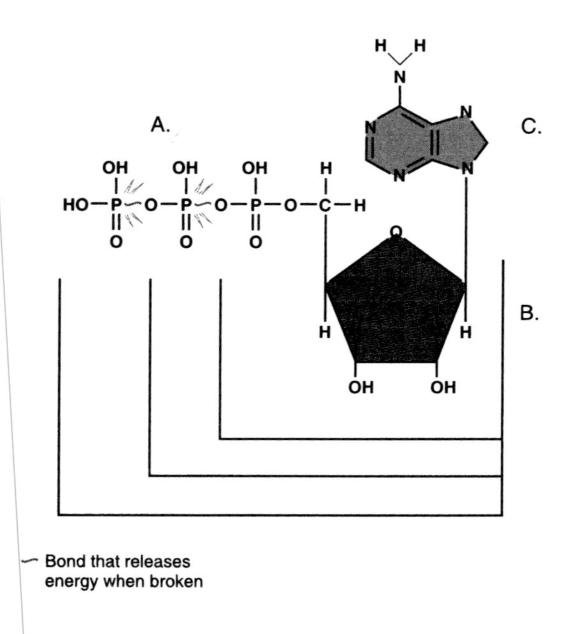

22. Which yields more energy, anaerobic respiration or aerobic respiration?

Which requires oxygen?

23. For every glucose molecule burned, the cell needs __________ oxygen molecules, and produces ____________ molecules of carbon dioxide and ____________ molecules of water. The maximum yield for aerobic respiration is ______ATP.

24. What is the final electron acceptor in aerobic respiration?_________________

25. Does the TCA cycle reduce or oxidize glucose? ________________

26. Glycolysis starts with __________________ and ends with _____________.

How many ATP molecules are generated in glycolysis for each molecule of glucose consumed?

27. How many carbons are in a glucose molecule? __________

How many carbons are in a pyruvic acid molecule? ________

How many pyruvic acids are produced for every glucose molecule metabolized?________

28. The TCA cycle produces __________.

Where do NADH molecules go with their electrons?

29. Which stage of glucose metabolism requires a membrane?

Why?

30. How does ATP synthase generate ATP?

31. Draw an ATP synthase molecule, a membrane, the H+ gradient, the flow of H+ ions, and the formation of ATP from ADP and P.

32. Why do we consider pyruvic acid a central part of metabolism?

What can pyruvate be converted to anaerobically?

Do all bacteria use pyruvate in the same way?

33. Where is most ATP generated in aerobic respiration?

34. How do fermentation and anaerobic respiration differ?

    Which yields more energy per glucose molecule?

35. How do alcoholic and acidic fermentation differ?

    Which fermentation do you want if you are making bread or beer?

    Which process sours milk?

    Which process do you want if you are making yogurt?

36. Amino acids are made up of carbon and nitrogen. Where can cells get the carbon?

    What do amino acids combine to form?

37. How are carbohydrates produced?

    Where are carbohydrates used in a bacterial cell?

38. How are lipids (fats) made?

    What are they used for in a procaryotic cell?

39. Many metabolic pathways are amphibolic. What does this mean?

40. Do precursor molecules (amino acids, sugars, lipids. come from the electron transport chain (Yes or No)?

41. If we labeled a glucose molecule's carbon atoms radioactively, so they could be traced, where would the carbons exit the metabolic pathway?

Organizing Your Knowledge

Table 1

Part of Aerobic Respiration	Location	Starting Molecules	End Products
Glycolysis	Cytosol	Glucose	a.
TCA cycle	Cytosol	b.	c.
Electron transport chain	d.	Loaded electron carriers, oxygen	ATP and water

Table 2

Metabolic Mechanism	Pathways Included	Final Electron Acceptor	Products	Microbes Using This
Aerobic respiration				
Anaerobic fermentation				
Anaerobic respiration				

Self-Test: Vocabulary

1._______reactions that breakdown larger molecules
2._______chemicals that increase the rate of a chemical reaction without being changed themselves
3. ______ the resistance to a reaction which must be overcome for a reaction to move forward
4. ______ reactant molecule upon which an enzyme works
5. ______ organic molecules that bind to an apoenzyme to form a holoenzyme
6._______enzymes that work outside a cell
7. ______ enzymes that are always present in a cell
8. ______ the loss of electrons
9. ______ chemically unstable
10. ______ loss of protein structure and enzyme function
11. ______ form of inhibition where an inhibitor binds to a regulator site to inhibit enzyme activity
12. ______ chemical reaction requiring energy to move forward
13. ______ the "energy currency" of a cell
14._______use of pyruvate or other organic molecule as an electron acceptor
15. ______ final electron acceptor in aerobic respiration
16. ______ process of central metabolism that requires a membrane
17. ______ type of fermentation desired in wine making
18. ______ during the TCA cycle, pyruvate is oxidized to ____
19. ______ theory describing the proton motive force fueling the synthesis of ATP
20. ______ integrating anabolic and catabolic pathways to improve efficiency in a cell

A. acidic
B. ADP
C. alcoholic
D. amphibolism
E. ATP
F. carbon dioxide
G. catabolism
H. catalysts
I. chemiosmosis
J. coenzymes
K. competitive
L. constitutive
M. denaturation
N. electron transport chain
O. endergonic
P. energy of activation
Q. exergonic
R. exoenzymes
S. fermentation
T. labile
U. noncompetitive
V. oxidation
W. oxygen
X. reduction
Y. substrates
Z. TCA cycle

Self-Test: Multiple Choice

1. Enzymes ____.

    a. add energy to chemical reactions
    b. increase the rate of chemical reactions
    c. are changed by the chemical reactions they catalyze
    d. work on all chemical reactions the same way

2. What is the final electron acceptor in aerobic respiration?

    a. oxygen
    b. carbon dioxide
    c. sulfur
    d. NADH

3. Which of the following factors will change enzyme function?

    a. temperature
    b. pH
    c. substrate concentration
    d. all of the above

4. An enzyme inhibitor that binds to the site normally used by a substrate and blocks enzyme function is called a ________.

    a. positive feedback inhibitor
    b. competitive inhibitor
    c. allosteric inhibitor
    d. enzyme inducer

5. The energy of activation of a chemical reaction ____.

    a. increases when enzymes are present
    b. decreases when enzymes are present
    c. is not changed by enzymes

6. Beta-galactosidase is not produced by a cell unless its substrate, lactose, is present. It metabolizes lactose inside the cell. We would describe this as a(n) ____ enzyme.

    a. constitutive endoenzyme
    b. induced endoenzyme
    c. induced exoenzyme
    d. constitutive exoenzyme

7. Enzyme cofactors are ____.

    a. generally vitamins and used to support enzyme function
    b. generally apoenzymes and work alone
    c. generally metallic and activate enzymes
    d. not used in bacterial cells, procaryotes have simple enzymes

8. If you labeled the carbons of glucose and sent it through aerobic respiration, where and how would the carbons be released?

    a. in glycolysis as carbon dioxide
    b. in glycolysis as water
    c. in the TCA cycle as water
    d. in the TCA cycle as carbon dioxide

9. Which portion of aerobic respiration requires a membrane to generate energy?

    a. glycolysis
    b. TCA cycle
    c. electron transport chain
    d. fermentation

10. Which part of central metabolism does NOT contribute precursor molecules to anabolic pathways?

    a. TCA cycle
    b. electron transport chain
    c. glycolysis
    d. pyruvic acid

11. The loss of electrons is a(n) ___.

    a. reduction
    b. oxidization
    c. condensation
    d. induction

12. The addition of water to chemical bonds ____ them and is a(n) ____ reaction.

    a. creates: anabolic
    b. breaks: anabolic
    c. creates: catabolic
    d. breaks: catabolic

13. Anabolic reactions ____ energy and are used in a cell for ____ reactions.

    a. release: synthesis
    b. use: degradative
    c. release: degradative
    d. use: synthesis

14. ___ is the energy currency of cellular reactions.

    a. DNA
    b. phosphate
    c. ATP
    d. AMP

15. Where is most of the energy (ATP. generated during aerobic respiration?

    a. glycolysis
    b. TCA cycle
    c. fermentation
    d. electron transport chain

Applications of Chapter 8

1. Most scientists agree that the first life on ancient Earth was anaerobic bacteria. Why do you think this is the case?

2. Microbial ecologists studying bioremediation are looking for microbes that are capable of degrading pollutants such as gasoline found in soils. From a metabolic point of view, what traits would you look for in such an organism?

Chapter 9 Microbial Genetics

Building Your Knowledge

1. The study of genetics explores many different subtopics. List four of these subtopics that geneticists study.

   a. ______________________________

   b. ______________________________

   c. ______________________________

   d. ______________________________

2. What is a genome?

   Give examples of four different ways that genetic material may be arranged in a cell.

3. What is a gene (describe several different definitions)?

4. What are the three basic categories of genes?

5. Is DNA longer or shorter than the cells that hold it?

6. What are the three basic components of DNA?

   Draw a nucleotide, labeling the three parts (deoxyribose, nitrogenous base, phosphate). Number the carbons on the deoxyribose 1–5.

7. What are complementary bases?

Do purines bind with each other or with pyrimidines?

What type of bonds hold complementary bases together?

What are the complementary base pairs in DNA? A bonds with ______ and C binds with _____. Are they the same in RNA?

8. What bonds hold the DNA backbone together?

Are these stronger or weaker than the bonds that hold the complementary bases together?

9. Draw a double-stranded DNA molecule, six nucleotides long, with complementary bases and a number of hydrogen bonds for each indicated, and label the orientation of each strand (3' and 5').

10. What does it mean that the strands of DNA are antiparallel to one another?

11. Why is it important that the nitrogenous bases in DNA have complementary pairs and that there are two different purines and two different pyrimidines (give 2 reasons)?

a.____________________________________

b.____________________________________

12. What enzymes are required for DNA replication and what is the function of each?

Enzyme Required	**Function**
Helicase	c.
a.	Synthesizes primer
Ligase	d.
b.	Supercoiling

13. If a bacterial cell was deficient in DNA polymerase I, would you expect greater or fewer mutations? ______________________

Explain your answer.

14. Draw the replication bubble of bacterial DNA replication; include in your drawing the origin of replication and all enzymes and factors required for replication.

15. Does RNA polymerase require a 3' OH to produce RNA from DNA? _________

Why is this important to DNA synthesis?

16. Why is DNA synthesized with a leading and a lagging strand?

17. Draw the flow of genetic information in cells that outlines how DNA guides the production of proteins.

18. How are DNA structure and protein function connected?

19. How is RNA different from DNA? Give three specific reasons.

a. ______________________________

b. ______________________________

c. ______________________________

20. What are the three stages of transcription and what happens during each stage?

21. What are the three types of RNA found in a cell, what is the function of each and are these translated?

RNA	Function	Translated?
mRNA	c.	Yes
a.	Carries amino acids	e.
b.	d.	f.

22. Draw the major steps in transcription. Label the coding strand, template strand of DNA, RNA polymerase, direction of transcription, and the growing mRNA transcript.

23. What are the three stages of translation?

a. ______________________________________

b. ______________________________________

c. ______________________________________

24. What is a codon?

25. Why is the genetic code of mRNA codons considered nearly universal? What are the exceptions to the “universal code”?

26. Draw a ribosome with an mRNA molecule and its A and P sites filled.

27. Given the RNA sequence AUG UUA CUA CCG GCG UAG, what would the amino acid sequence look like?

28. What is a promoter?

29. What happens when a ribosome encounters a nonsense codon on an mRNA message?

30. What is a polyribosomal complex?

Do these complexes exist in eucaryotes?

31. What are introns and exons?

Do bacterial cells have introns and exons?

32. Viruses can be either DNA or RNA viruses. Which are generally double-stranded?

Where are double-stranded DNA (dsDNA) viruses replicated in animal cells?

How are dsDNA viruses replicated? (Where does the polymerase come from?)

33. Diagram the lac operon.

a. What does it look like when lactose is not available?

b. What does it look like when lactose is available?

34. How is a repressible operon affected by a corepressor?

35. How may antibiotics affect transcription and translation in prokaryotes?

36. How do wild-type strains differ from mutant strains of a bacterial species?

37. What is the Ames test and what is it used for?

Do bacteria treated with mutagens show more or fewer colonies than those treated with nonmutagens when tested in the Ames test?

38. How do spontaneous mutations differ from induced mutations?

Do all organisms have the same mutation rate in nature?

39. Compare and contrast the actions of ethidium bromide, nitrogenous base analogs and X rays.

40. Compare and contrast same-sense, missense and nonsense mutations.

41. Frame-shift and nonsense mutations often knockout expression of a particular gene. Why do you think this is the case?

42. What is genetic recombination in bacteria?

43. How may genes be transferred from one bacterial cell to another? List and define three processes.

Transfer Process	**Definition**
Transformation	a.
b.	Joining of two bacteria and transfer of DNA
Transduction	c.

44. Diagram the process of conjugation.

45. Differentiate between generalized and specialized transduction.

Generalized:______________________________

Specialized:______________________________

46. What are three ways a transposon can move?

Organizing Your Knowledge

Table 1

Nucleotide	Purine or Pyrimidine?	Pairs With	Purine or Pyrimidine?
Thymidine	a.	c.	e.
Guanine	b.	d.	f.

Table 2

Template	Product	Enzyme	Process
DNA	RNA	a.	e.
RNA	DNA	b.	f.
mRNA	Protein	c.	g.
DNA	DNA	d.	h.

Table 3

Enzyme, Protein, or Factor	Purpose	Process It's Involved In
Helicase	a.	Replication
RNA primer	Provides 3'OH	h.
Primase	b.	Replication
DNA polymerase I	Proofreads DNA	i.
DNA polymerase III	c.	Replication
Ligase	Joins DNA fragments	j.
Gyrase	d.	Replication
RNA polymerase	Synthesizes RNA	k.
Codons	e.	Translation
Polyribosomal complex	Rapid translation	l.
Operon	f.	Translation
mRNA	Transmits message	m.
rRNA	g.	Translation
tRNA	Carries amino acids	n.

Self-Test: Vocabulary

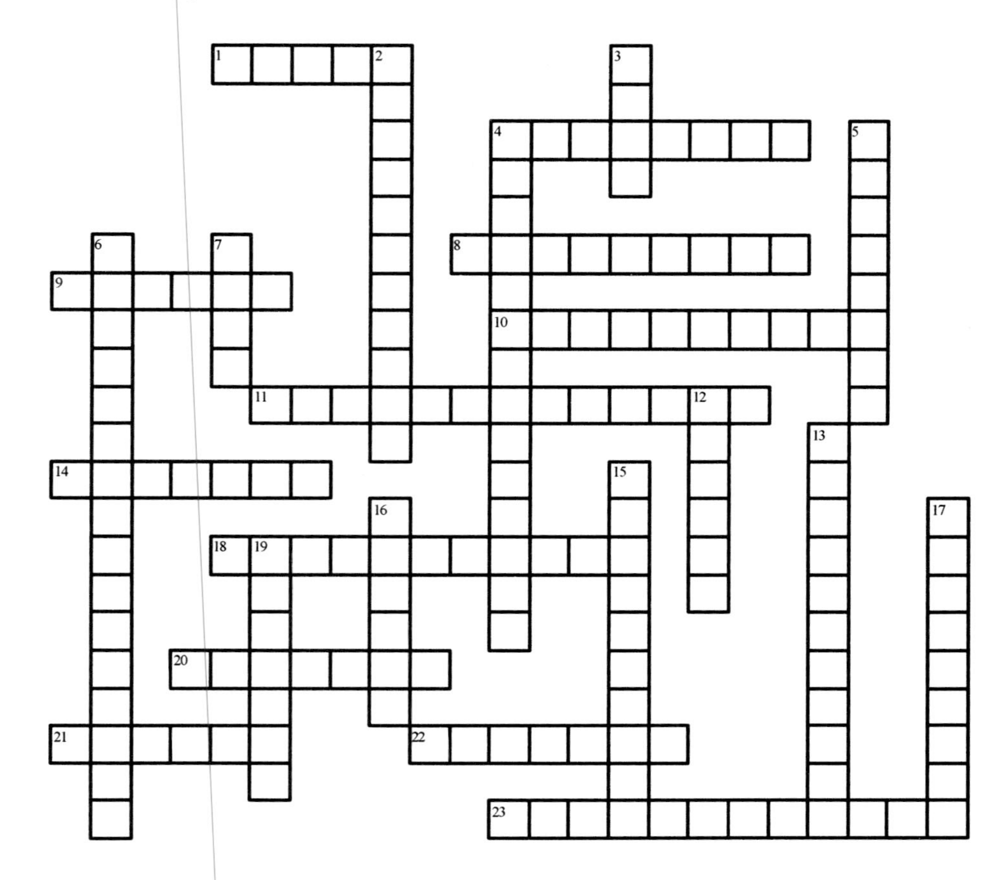

ACROSS

1. mutation that is the addition, deletion, or substitution of a single base
4. form of RNA that carries amino acids
8. triplet found in tRNA that is the complement to the codon found in mRNA
9. sum of the genetic material in an organism
10. mutation that changes the reading frame of a gene
11. use of DNA template to make RNA
14. mutations resulting from exposure to known mutagens
18. type of transduction that requires a prophage
20. enzyme that makes an RNA primer
21. enzyme that seals nicks in DNA and joins Okazaki fragments
22. sequences that are in a gene but not translated
23. complex of multiple ribosomes binding to a single mRNA

DOWN

2. jumping genes
3. unit of heredity that determines a trait
4. form of DNA transfer that does not require viruses or physical contact
5. region of DNA recognized by RNA polymerase to start transcription
6. model that describes and explains how DNA is replicated with reliability and fidelity
7. test that can be used to screen chemicals for carcinogenic potential
12. collection of genes that are regulated as a single unit
13. purines and pyrimidines are both this type of base
15. reason why mRNA structure cannot be determined by amino acid sequence
16. five-carbon sugar found in RNA
17. expression of traits or the appearance of an organism
19. adenine and guanine

Self-Test: Multiple Choice

1. Replication is ________.
   a. The production of DNA from a DNA template.
   b. The production of RNA from a DNA template
   c. The production of RNA from a RNA template
   d. The production of DNA from a RNA template

2. Okazaki fragments are __.
   a. the normal result of leading strand synthesis
   b. the result of mistakes in transcription
   c. the normal result of lagging strand synthesis
   d. the result of mistakes in translation

3. The sum total of the genetic material of a cell is called its ____.
   a. plamids
   b. chromosome
   c. polyribosomal complex
   d. genome

4. The addition of one or two bases in a DNA message will often lead to a ____.
   a. translocation
   b. inversion
   c. frame shift
   d. compensatory deletion

5. The complementary base pairs of DNA are held together by___.
   a. covalent bonds
   b. hydrogen bonds
   c. ionic bonds
   d. ionocovalent bonds

6. RNA polymerase binds to ____.
   a. the DNA of the promoter region
   b. the RNA of the promoter region
   c. the DNA of the start codon
   d. the RNA of the start codon

7. The formation of a polyribosomal complex is _____.
   a. part of normal DNA replication
   b. part of abnormal RNA transcription
   c. a nonsense mutation
   d. a part of normal translation

8. Which of the following is NOT a form of RNA found in cells?
   a. ribosomal RNA
   b. transfer RNA
   c. messenger RNA
   d. lysosomal RNA

9. Which of the following nucleotides are purines?
   a. adenine and thymine
   b. cytosine and guanine
   c. adenine and guanine
   d. cytosine and adenine

10. Which method of gene transfer between procaryotes requires DIRECT contact?
    a. transformation
    b. transduction
    c. conjugation
    d. all of the above require direct contact between donor and host

11. The genetic code is called universal because _______.
    a. all procaryotes use the same code
    b. all eucaryotes use the same code
    c. all mammals use the same code
    d. all living things use the same code

12. Generalized transduction does NOT require ____.
    a. phage
    b. transfer of genetic material
    c. DNA
    d. prophage

13. Uncorrected errors in DNA replication become___.
    a. replication forks
    b. mutations
    c. lagging strands
    d. gyrases

14. Which of the following statements is FALSE, concerning DNA?
    a. The two strands of DNA are in antiparallel orientation to each other.
    b. The backbone of DNA is made up of sugars and phosphates linked by ionic bonds.
    c. The rungs of the DNA ladder are made up of complementary base pairs.
    d. DNA replication is semiconservative.

15. If you see a sequence of single-stranded nucleic acid with uracil and no thymine, you are looking at ____.
    a. plasmid DNA
    b. chromosomal DNA
    c. RNA, not DNA
    d. Okazaki fragments

Applications of Chapter 9

1. The Ames test often includes test groups where the potential mutagen is exposed to liver enzymes. Why would researchers add liver enzymes to a compound before testing it in the Ames test?

2. How does genetic recombination between bacteria facilitate the rapid spread of resistance to antimicrobial agents?

Chapter 10 Genetic Engineering

Building Your Knowledge

1. What is genetic engineering?

2. Give two examples of how we can use genetic engineering in society.

   a. ____________________________

   b. ____________________________

3. How do scientists cut DNA?

   What types of sequences do these "molecular scissors" recognize?

   Why is the creation of "sticky ends" important when cutting DNA?

4. How is ligase used in genetic engineering?

5. What is cDNA and how is it made?

   Does cDNA have introns?

6. What is gel electrophoresis?

   Why do DNA molecules move toward the positive pole of a gel?

   Which move faster in a gel, large pieces of DNA or small pieces?

   Are the slow-moving pieces found at the top (negative end) or the bottom (positive end) of the gel?

7. What are gene probes and what can they be used for?

8. Diagram the process of Sanger DNA sequencing. How many tubes do you set up? Why?

   Why do dideoxynucleotides stop the elongation of DNA?

9. If you were going to set up a test tube for PCR, what would you add?

10. Why is it important that you use a thermostable enzyme for PCR?

11. What is recombinant DNA?

12. Diagram the process of cloning. Label the insert DNA, vector, and cloning host and tell how we identify clones with the gene of interest.

13. Cloning vectors must have three key traits. Name them.

14. What are some of the desirable features of cloning hosts?

15. Why are recombinant proteins, such as insulin, better medicines than animal- or human-derived proteins? What other recombinant proteins have been produced?

16. What are transgenic organisms?

17. Give three examples of genetically engineered plants, listing the gene inserted and the advantage of having the new gene.

Plant	**Gene Inserted**	**Advantage**
a.		
b.		
c.		

18. Give four examples of genetically engineered animals, listing the gene alterations and the reason for these alterations.

Animal	**Gene Alteration**	**Reason for Alteration**
a.		
b.		
c.		
d.		

19. What is gene therapy and how does it differ from recombinant protein production?

20. Diagram the process of DNA fingerprinting. How can DNA fingerprinting be used? (List multiple ways.)

21. How has PCR improved the use of DNA fingerprinting by forensic scientists?

22. What do DNA microarrays identify?

Organizing Your Knowledge

Table 1. Nature vs. Lab

Enzyme	**Use in Nature**	**Use in Lab**
Restriction endonucleases	Defense against viral attack	b.
Ligase	a.	Joining cloned fragments
Reverse transcriptase	Retroviral cycles	c.

Table 2. Methods of Genetic Engineering

Method	**Purpose**
Sanger sequencing	a.
Southern blot	b.
Transgenic organisms	c.
PCR	d.
Microarrays	e.
DNA fingerprinting	f.
Cloning	g.
Hybridization tests	h.

Self-Test: Vocabulary

1. ______ study that relates the science of biology to moral judgment of human conduct
2. ______ direct, deliberate manipulation of an organism's genome
3. ______ enzymes used to cut DNA at specific palindromic sites
4. ______ sequence of DNA that reads the same 5'–3' and 3'–5'
5. ______ enzyme that seals "sticky ends"
6. ______ enzyme that makes DNA from an RNA template
7. ______ procedure in which DNA fragments are placed in an electrical field and separated based on size
8. ______ short, labeled DNA fragments that bind to complementary (unlabeled) fragments
9. ______ test where DNA from a sample is denatured, placed on a filter then screened with a specific probe
10. ______ process that provides the identity and order of nucleotides in a DNA fragment
11. ______ technique that rapidly amplifies the amount of DNA in a sample
12. ______ short segments of DNA that bind to DNA at specific regions during the priming phase of PCR
13. ______ heating DNA to separate it into two strands
14. ______ a plasmid or virus that is used to carry DNA into a cloning host
15. ______ recombinant organisms produced by adding foreign genes to their genomes
16. ______ use of transgenic animals or plants to produce medically useful products
17. ______ analysis of DNA to emphasize the difference between individual DNA sequences
18. ______ observable differences in DNA structure

A. bioethics
B. denaturation
C. DNA fingerprinting
D. DNA sequencing
E. gel electrophoresis
F. gene probes
G. genetic engineering
H. hybridization test
I. ligase
J. markers
K. palindrome
L. pharming
M. polymerase chain reaction
N. primers
O. priming
P. restriction endonucleases
Q. reverse transcriptase
R. RNA polymerase
S. transgenic
T. vector

Self-Test: Multiple Choice

1. When you run an electrophoresis gel and see several bands of DNA, the larger pieces of DNA are found ____.
   a. at the top because they move faster
   b. at the bottom because they move slower
   c. at the top because they move slower
   d. at the bottom because they move faster

2. Which of the following would you NOT add to a PCR reaction?
   a. thermostable polymerase
   b. template
   c. primers
   d. ligase

3. Ligase is used by molecular biologists to___.
   a. cut DNA fragments
   b. join DNA fragments
   c. convert RNA to DNA
   d. drive plasmids out of solution

4. Organisms are transgenic if they ___.
   a. are hybrids from two species —like mules
   b. lack chromosomes
   c. have a foreign gene inserted into them
   d. lack all genes

5. Restriction endonucleases recognize __.
   a. mismatched DNA segments and repair them
   b. palindromic DNA sequences and cut them
   c. mismatched DNA sequences and cut them
   d. palindromic DNA sequences and repair them

6. A plasmid is ___________.
   a. found only in fungus
   b. a bit of chromosomal DNA
   c. linear DNA found in prokaryotes
   d. circular DNA separate from chromosomal DNA

7. DNA fingerprinting has been used to _____.
   a. identify individuals after a disaster
   b. identify microbes
   c. determine parentage
   d. all of the above

8. Cloning of DNA fragments and inserting them into a host forms ____.
   a. vectors
   b. cloning hosts
   c. recombinant organisms
   d. PCR reactions

9. PCR is used to ________ .
   a. amplify DNA segments
   b. cut DNA segments
   c. join DNA segments
   d. force DNA segments into host cells

10. Antisense DNA will bind to ___ and prevent translation.
    a. tRNA
    b. mRNA
    c. ribosomes
    d. rRNA

11. Recombinant human proteins are often better than animal products or human products because ___.
    a. infectious diseases can be spread through these proteins
    b. humans often develop allergies to animal proteins
    c. there often isn't enough protein to treat all the patients that need a given protein
    d. all of the above

12. The steps of the PCR cycle are ____ (in order).
    a. denaturation, priming, extension
    b. cutting, denaturation, priming
    c. priming, denaturation, cutting
    d. extension, denaturation, priming

13. Dideoxynucleotides are used for ______.
    a. cutting DNA
    b. joining DNA fragments
    c. amplifying DNA
    d. sequencing DNA

14. cDNA is made by ____.
    a. incubating DNA with RNA polymerase
    b. combining plasmid and target DNA
    c. incubating RNA with reverse transcriptase
    d. cutting DNA probes

15. Good cloning hosts need all of the following traits EXCEPT ______.
    a. fast growth rate
    b. well-known genome
    c. pathogenicity
    d. maintenance of foreign DNA through multiple generations

Applications of Chapter 10

1. Modern gene therapy protocols have been approved that affect human somatic cells, but not germ-line cells (eggs and sperm). Why do you think this is the case?

2. Gene therapy can be used to treat several different single gene defects, but not all. What would prevent all genetic defects from being treated with gene therapy?

Chapter 11 Physical and Chemical Control of Microbes

Building Your Knowledge

1. What are contaminants?

2. What are the three general methods used in decontamination protocols?

3. What is the difference between physical and chemical methods of microbial control?

4. Are all contaminants equally susceptible to physical and chemical means of decontamination?

5. How do sterilization and disinfection differ?

   Can something be almost sterile? Explain.

6. An antimicrobial agent can be -cidal or -static (e.g., bactericidal or fungistatic). What is the difference between a -cidal agent and a -static agent?

   If you take bacteriostatic antibiotics for an infection, what will happen when you stop taking the antibiotic?

7. Differentiate between the use and definition of antiseptics and disinfectants.

Agent	Used On	Examples
Disinfectant	a.	c.
Antiseptic	b.	d.

8. Your bathroom in a public restroom may say "sanitized for your protection." Is sanitization the same as sterilization?

9. When is a microbe considered dead?

10. How do the number of organisms and their state of growth (spores vs. vegetative cells) affect microbial death rate?

11. How does a chemical agent's concentration affect microbial death rate?

12. What are the four major cellular targets for antimicrobial agents?

    a. ______________________________

    b. ______________________________

    c. ______________________________

    d. ______________________________

13. Name one agent that targets bacterial cell walls.

14. How do surfactants work?

15. How may proteins be denatured?

16. How does heat affect microbial cells? Which is more effective and why?

    a. Moist heat: ___________________________

    b. Dry heat: ______________________________

17. If organism A has a TDT of 15 minutes and organism B has a TDT of 35 minutes, using the same temperature, which organism is most resistant to heat?

18. Why is pressure added to the steam present in an autoclave?

19. What is tyndallization?

20. Does boiling water sterilize the water?

21. Does pasteurization sterilize milk?

What does it do?

22. Why is cold not used to kill microbes?

23. Why does radiation limit microbe growth?

a. Ionizing radiation: ____________________

b. Nonionizing radiation: ________________

24. How is the usefulness of nonionizing radiation limited?

25. When is filtration an effective method to remove contaminants?

26. How do aqueous solutions differ from tinctures?

27. Describe several characteristics that the "ideal" chemical antimicrobial agent would have.

28. How does bleach work as a disinfectant?

29. How does alcohol kill bacteria?

30. What is the cellular target of detergents?

31. What are the cellular targets of heavy metals?

32. Glutaraldehyde is used to sterilize contaminated areas—how does it kill microbes?

33. Ethylene oxide is a sterilant. Why is it not used on human tissues?

Organizing Your Knowledge

Control Method	Type (Physical, Chemical, Mechanical)	Cellular Target	Example
Moist heat	a.	Proteins (denaturant)	Autoclave
Dry heat	Physical	Proteins (dehydration, oxidation)	m.
Ethylene oxide	Chemical	g.	Chemiclave
Pasteurization	b.	Denatures proteins	Pasteurizing milk
Halogenation	Chemical	Disrupts S bond-S bonds	n.
Filtration	Mechanical	h.	HEPA filters
Ionizing radiation	c.	Breaks DNA	Irradiation of mail
Nonionizing radiation	Physical	Induces mutations	o.
Surfactants	Chemical	i.	Rubbing alcohol
Hydrogen peroxide	d.	Oxidizes proteins	Antiseptic use
Quaternary ammonia compounds	Chemical	Detergent	p.
Heavy metals	Chemical	j.	Thimerosal
Aldehydes	e.	Cross-link proteins	Cidex
Sanitizing	Mechanical	k.	q.
Cold temperatures	Physical	l.	Refrigeration
Phenolics	f.	Protein denaturant	Triclosan (soap)

Self-Test: Vocabulary

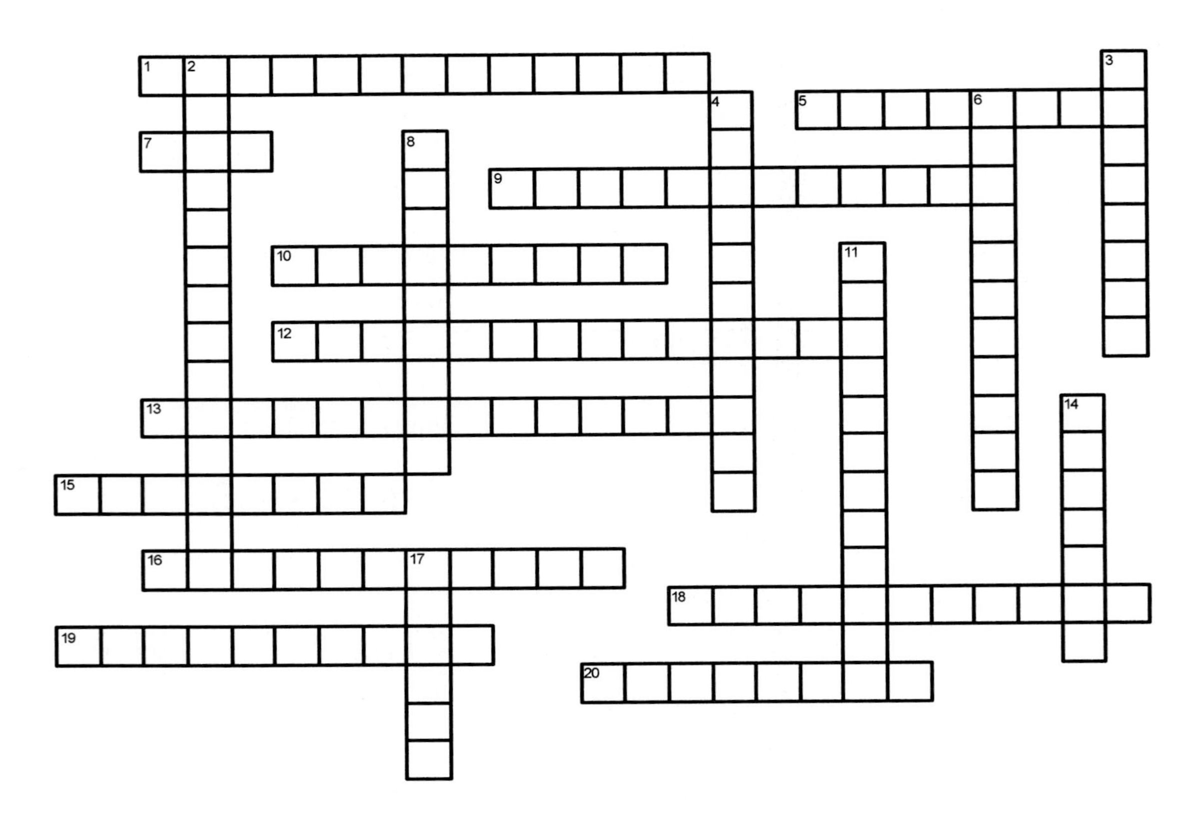

ACROSS

1 process that removes or destroys all microbes (including endospores)
5 disrupt of protein structure and function
7 cellular target of ionizing and non-ionizing radiation
9 chemical that kills bacteria
10 device that sterilizes objects by exposing them to steam under pressure
12 preserving by freezing and drying
13 applying heat to liquids to remove spoilage and infectious agents
15 radiation that breaks the sugar-phosphate backbone of DNA
16 chemical agents that are used directly on living tissues to decontaminate
18 agent that stops fungal growth
19 to become dehydrated
20 chlorine and iodine for example

DOWN

2 intermittent sterilization
3 cellular target of detergents
4 microbicidal agents that lower the surface tension of the membrane and disrupt it
6 ability to withstand heat
8 bacterial form that is the most resistant to microbial control methods
11 cleansing technique that removes debris, microbes and toxins
14 practices that prevent entry of microbes into sterile tissues and prevent infection
17 compound all chemicals are compared against when rating their effectiveness.

Self-Test: Multiple Choice

1. Cooking utensils are often ____ between uses in restaurants.
   a. sanitized
   b. irradiated
   c. sterilized
   d. autoclaved
2. Dry heat is more effective at killing microbes than moist heat
   a. True
   b. False
3. Which of the following is NOT a physical agent used to control microbes?
   a. moist heat
   b. dry heat
   c. radiation
   d. ethylene oxide
4. 100% Ethanol solutions are more effective antimicrobial agents than are 70% solutions.
   a. True
   b. False
5. Alcohol sterilizes skin.
   a. True
   b. False
6. Nonionizing radiation, such as ____ kills bacteria by ____.
   a. UV: denaturing lipids
   b. X rays: denaturing proteins
   c. UV: inducing mutations in DNA
   d. X rays: inducing mutations in RNA
7. Household bleach is used as a common disinfectant because it has ____.
   a. hypochlorite
   b. idonic
   c. phenol
   d. chlorohexidine
8. Placing objects in boiling water _____ them.
   a. sterilizes
   b. disinfects
   c. sanitizes
   d. tyndallizes
9. Which of the following has the HIGHEST resistance to killing?
   a. naked viruses
   b. vegetative bacterial cells
   c. bacterial endospores
   d. viral endospores
10. A chemiclave sterilizes objects by exposing them to _____ gas.
    a. formaldehyde
    b. surfactant
    c. ethylene oxide
    d. silver tincture
11. Which of the following is NOT a chemical agent used to control microbes?
    a. filtration
    b. phenol
    c. halogens
    d. glutaraldehyde
12. An autoclave sterilizes objects by using ____.
    a. chemical treatments
    b. multiple low-heat treatments
    c. steam under pressure
    d. multiple incinerators
13. Chemical surfactants kill microbes by ___.
    a. damaging their DNA
    b. stopping ribosomal movement
    c. denaturing proteins
    d. disrupting membranes
14. A viricidal agent will ___.
    a. kill bacteria
    b. kill viruses
    c. stop bacterial growth, but not kill the bacteria
    d. stop viral growth, but not kill the viruses
15. Treatment with cold ____.
    a. slows microbial growth
    b. kills all microbes
    c. can be used to disinfect surfaces
    d. kills all fungus

Applications of Chapter 11

1. If you had a nonautoclave safe object that you needed to use as a surgical implant, how would you clean it for use? Discuss several options and advantages of each option.

Chapter 12 Drugs, Microbes, and Host Elements of Chemotherapy

Building Your Knowledge

1. One hundred years ago, what percentage of all children in the United States died of infectious disease before age 5?

2. Design the perfect antibiotic.

   Does the perfect antibiotic actually exist? Why or why not?

3. Where do most antibiotics originate, in the lab or from nature?

   Why is this the case?

4. Differentiate between chemotherapeutic and prophylactic drug use.

   Is taking penicillin for a case of Strep throat prophylactic or chemotherapeutic?

5. Explain the concept of selective toxicity.

6. Which drugs are most selectively toxic to bacterial cells (in general)?

   Which are the least selectively toxic?

7. Are drugs that target the cell wall more or less selectively toxic than those that target the plasma membrane? Why?

8. Why are some drugs broad-spectrum and others narrow-spectrum?

9. Are penicillin and penicillin-like antibiotics more effective against actively growing cells, or old, dormant cells? Why?

10. Are the membrane-disrupting drugs generally used topically (on body surfaces) or administered internally?

11. How are chloroquine and AZT similar?

    How are they different?

12. If both eucaryotes and procaryotes have ribosomes, why are antimicrobials that target ribosomes selectively toxic?

13. What is competitive inhibition and what does it have to do with sulfonamide activity against bacterial cells?

14. Fill in the diagram with several examples of antibiotics that target each of the following structures or processes in a bacterial cell.

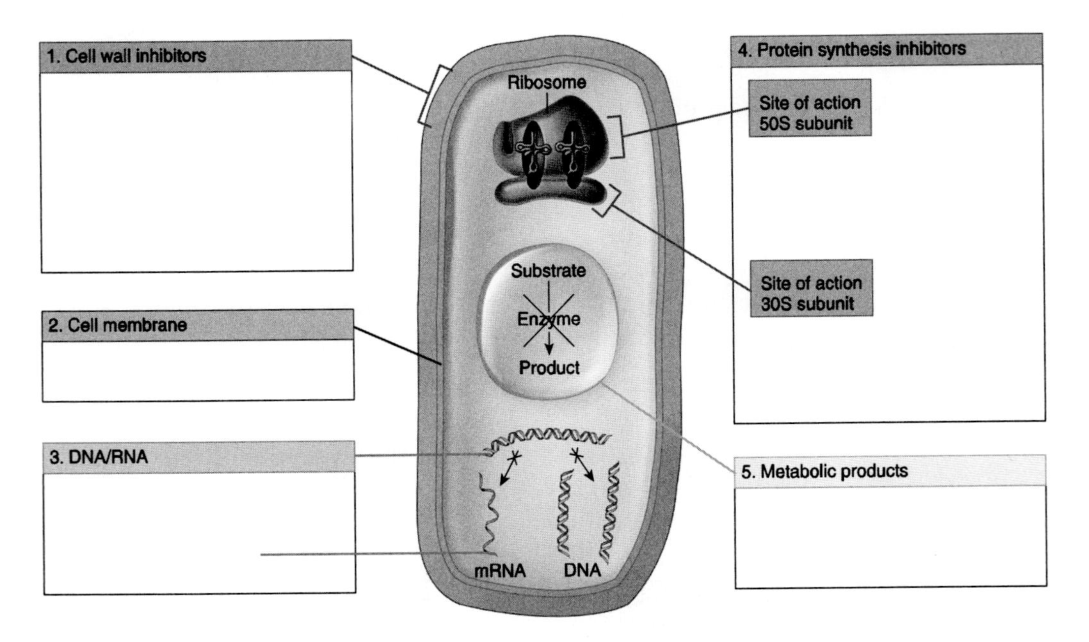

15. How is most penicillin produced?

16. List three members of the penicillin family.

a. ______________________________

b. ______________________________

c. ______________________________

17. What is the advantage of using semisynthetic penicillins, like ampicillin or nafcillin?

18. What are the major problems that limit the usefulness of the penicillin antimicrobials?

19. Why add clavulanic acid to penicillins (e.g., Augmentin)?

20. What are cephalosporins? What are the "generations" of cephalosporins based on?

How are they generally administered? Why?

21. What is the mechanism of action of the fluoroquinolones?

22. What are the aminoglycosides and what infections are they used to treat?

23. What is the mechanism of action of tetracycline?

    Why is tetracycline not as commonly prescribed as amoxicillin?

24. Why is chloramphenicol not a widely used antimicrobial?

25. What do erythromycin and clindamycin have in common?

    How do these drugs act against bacteria?

    Why is clindamycin not widely used?

26. Why are scientists hopeful that resistance to the oxazolidinones will be slow to develop?

27. Where do sulfa drugs come from?

    How does this origin differ from that of penicillins and cephalosporins?

28. What are fosofmycin and synercid?

    Why are they not widely used?

29. Why are antifungals generally more toxic to human tissues than antibacterial agents?

30. Describe two different antifungal agents; include in your description the conditions they treat.

Why are polyenes effective against fungal cells, but not bacterial cells?

31. Why are there few effective antiparasite drugs?

What drugs are used to treat malaria?

What drugs are used to treat roundworm?

32. Why is selective toxicity so difficult to achieve in antiviral therapies?

33. Why are viral diseases like measles and mumps fairly rare in the United States?

34. What are three basic mechanisms of action of antiviral agents?

a.____

b.____

c.____

35. Why is the fact HIV is a retrovirus significant when designing antiviral therapies?

36. What drugs are commonly used to treat HIV and how do they inhibit the viral cycle?

37. How does chromosomal drug resistance originate? Does this type of resistance spread in a population?

38. What are R factors? Do these spread through a bacterial population?

39. Describe four distinct ways bacteria may become resistant to antibiotics they were once sensitive to.

a.____

b.____

c.____

d.____

40. What are beta-lactamases and what antimicrobial drugs do they confer resistance to?

41. Do pumps generally confer resistance to one type of antimicrobial or many? Why?

42. How do bacteria become resistant to sulfonamide?

43. Does exposure to an antibiotic increase or decrease the percentage of resistant cells in a population? Explain.

44. Why does combining drug therapies limit the spread of drug resistance?

45. There are three major categories of antibiotic side effects. Name them.

a. ____

b. ____

c. ____

46. Why are the liver and kidneys often damaged by antibiotics?

47. Why are tetracyclines not given to pregnant women or young children?

48. How may antibiotics cause diarrhea (list two ways)?

49. If a person takes penicillin once and does not have an allergic reaction to it, does that mean they are not allergic? Explain.

50. What three factors do doctors generally consider when choosing antimicrobial therapies?

a. ______________________________

b. ______________________________

c. ______________________________

51. What two methods are commonly used to tell which antibiotics are most and least effective against a given pathogen?

a. ______________________________

b. ______________________________

52. Diagram a Kirby-Bauer plate. Draw the plate, antibiotic disks, and zones of inhibition.

If drug A has a larger zone than drug B and both drugs are at the same concentration and are the same size, which drug, A or B, is more effective?

53. What is the MIC and how is it used?

54. What is the therapeutic index (TI)?

If a drug has a TI of 5 is it more or less safe to use than one with a TI of 0.5? Why?

55. What other variables do physicians need to be concerned with when choosing antimicrobial drugs?

Organizing Your Knowledge
Table 1

Antimicrobial Agent	**Mechanism of Action**	**Commonly Used to Treat**
Penicillin	Cell wall inhibitor	g.
Sulfonamide	a.	Otitis media, PCP, urinary tract infections
Chloroquine	Blocks RBC infection	h.
Gentamicin	b.	Gram-negative infections
Polymyxins	Disrupts cell membranes	i.
Nystatin	c.	Superficial fungal infections
Rifampicin	Inhibits RNA polymerase	j.
Metronidazole	d.	Protist infections
AZT	Nucleotide analog—blocks DNA synthesis	k.
Amphotericin B	e.	Systemic fungal infections
Chloramphenicol	Inhibits peptide bond formation	l.
Tetracycline	f.	Lyme disease, mycoplasmas

Self-Test: Vocabulary

1._______ beta-lactamase inhibitor that is added to amoxicillin to form Augmentin

2._______ use of a drug to prevent infection

3._______ origins of sulfa drugs

4._______ drugs that are chemically modified in the lab after being isolated from nature

5._______ killing target microbes without harming the host

6._______ a drug that is useful against a small number of organisms (e.g., gram-negative only)

7._______ inhibitor of mycolic acid that is used to treat tuberculosis

8._______antimicrobial agents that target sterols in fungal membranes

9._______ metabolic pathway inhibited by sulfa drugs

10._______ class of antibiotics produced by actinomycetes

11._______ specific process inhibited by erythromycin

12._______ enzyme that confers resistance to penicillins and cephalosporins

13._______ process blocked by chloramphenicol

14._______ the first antimalarial drug

15._______ antiviral drug that terminates DNA replication in herpesviruses

16._______ plasmids that encode drug resistance genes

17._______ structures that pump antimicrobial agents out of a cell

18._______ overgrowth of resident organisms that causes disease

19._______ agar-diffusion test that provides information on antimicrobial sensitivity

20._______ synthetic drugs that bind to gyrase to inhibit DNA synthesis

A. acyclovir

B. aminoglycosides

C. aniline dyes

D. beta-lactamase

E. clavulanic acid

F. fluoroquinolones

G. folic acid synthesis

H. fourth generation

I. isoniazid (INH)

J. Kirby-Bauer

K. MDR pumps

L. narrow-spectrum

M. peptide bond formation

N. polyenes

O. prophylaxis

P. quinine

Q. R factors

R. selective toxicity

S. semisynthetic

T. superinfection

U. synthetic

V. translocation

Self-Test: Multiple Choice

1. Which of the following is NOT a common target for antibacterial drugs?
   a. cell wall synthesis
   b. nucleic acid structure
   c. protein synthesis
   d. bacterial cell nucleus

2. All of the following antibiotics target procaryotic ribosomes EXCEPT___.
   a. streptomycin
   b. cephalexin
   c. gentamicin
   d. erythromycin

3. Prophylatic antibiotics are given ___.
   a. after a person is infected with a virus
   b. to people at increased risk of viral infection
   c. after a person is infected with bacteria
   d. to people at increased risk of bacterial infection

4. An antibiotic with a high therapeutic index (TI) ___.
   a. is a less risky choice than one with a low TI
   b. is generally very toxic
   c. has a high MIC and low toxic dose

5. Which of the following is NOT a characteristic of an ideal antimicrobial drug?
   a. not excessive in cost
   b. microbistatic, not microbicidal
   c. selectively toxic to microbe
   d. easy to deliver to site of infection

6. Antiviral drugs are ___.
   a. commonly used to treat head colds
   b. hard to design because the viruses are intracellular parasites
   c. generally safer to use than anti-bacterial drugs
   d. not subject to antiviral resistance

7. An MDR pump will confer resistance to ___.
   a. a single class of antibiotics (e.g., the penicillins)
   b. many different antibiotics from different groups
   c. only gram-negative bacteria
   d. only gram-positive bacteria

8. ____ is an example of a synthetic antimicrobic drug.
   a. polymyxin
   b. rifamycin
   c. tetracyline
   d. sulfonamide

9. When the cause of a disease is unknown, but suspected to be bacterial, a useful course of action would be ____.
   a. to start antiviral therapy
   b. to disinfect the patient
   c. to start a broad-spectrum antibiotic
   d. to start a narrow-spectrum antibiotic

10. Two ways to determine an organism's resistance to antimicrobial drugs are ___ and ___ methods.
    a. MIC: therapeutic index
    b. Kirby-Bauer: therapeutic index
    c. MIC: Kirby-Bauer
    d. Kirby-Bauer: beta-lactamase

11. An organism becomes resistant to penicillin when it___.
    a. produces thymidine kinase
    b. acquires its folic acid from the enviroment
    c. produces beta-lactamase
    d. loses its DNA

12. Clavulanic acid is added to the penicillin group of drugs because ___.

    a. it inhibits beta-lactamase enzymes
    b. it works against gram-positive bacteria, penicillins don't
    c. it lengthens the shelf-life of penicillins
    d. it disrupts bacterial cell membranes

13. The major drawbacks to penicillin use are ____.

    a. development of resistance and host cell toxicity
    b. synergistic effects with antiviral therapies
    c. development of resistance and host allergic responses
    d. purine degradation and host allergic responses

14. Antibiotics that disrupt microbial plasma membranes ___.

    a. are more toxic than those that disrupt microbial cell walls
    b. are commonly given systemically
    c. are not toxic to host cells
    d. generally have a high therapeutic index

15. ____ is commonly used to treat fungal infections.

    a. Tetracycline
    b. Vancomycin
    c. Amphotericin B
    d. Quinine

Applications of Chapter 12

1. Antibiotics may be either broad-spectrum or narrow-spectrum. Despite the dangers of selecting for antibiotic resistance, broad-spectrum antibiotics are commonly the first antimicrobial drugs prescribed, especially in the hospital. Why do you think this is the case?

2. Several newer antimicrobial agents are either completely synthetic or significantly modified from their native form. What are some advantages and disadvantages to this approach of drug development?

Chapter 13 Microbe–Host Interactions: Infection and Disease

Building Your Knowledge

1. How do pathogens differ from normal flora?

2. What are the possible outcomes following microbial contact with a human host?

3. Once a person develops a disease, what can happen to the individual?

4. Where are most of the normal flora in the human body found?

   How do transient flora differ from resident flora?

5. In general, which sites of the human body lack normal flora?

6. Give two specific examples of how normal flora may protect against infection.

   a. ____________________________________

   b. ____________________________________

7. What is an endogenous infection? Which people are at particular risk for endogenous infections?

8. Describe how newborn infants acquire normal flora.

9. What conditions on the skin support the growth of normal flora?

   Which conditions inhibit the growth of normal flora?

10. Where are most of the normal flora found in the gastrointestinal tract?

    Why do doctors often recommend eating yogurt?

11. Does the entire respiratory tract have normal flora? Explain.

12. Bladder infections are often endogenous infections—are they more common in men or women?

13. How do true pathogens differ from opportunistic pathogens?

14. What are virulence factors?

15. Describe three separate virulence factors and how they contribute to the disease process.

    a. ______________________________

    b. ______________________________

    c. ______________________________

16. Which are more dangerous, pathogens at biosafety level 1 or biosafety level 4?

17. How is the portal of entry important to the spread of disease? If a person has influenza virus on his or her hands, will that person get influenza? Explain.

18. Describe three separate portals of entry and the pathogens that enter the body through them.

19. Define infectious dose. Would you rather drink 1000 *V. cholera* cells or inhale 10 measles viral particles? Why?

20. Why is adhesion of particular importance to pathogens?

    Describe three different adhesion methods used by pathogens.

21. How are an exoenzyme and exotoxin similar?

    How are they different?

    Give three specific examples of exoenzymes that are virulence traits.

22. Differentiate between an intoxication and infection. Which one requires active growth of bacteria?

23. Differentiate between exotoxins and endotoxins.

Give three specific examples of exotoxins as virulence traits.

a. ______________________________

b. ______________________________

c. ______________________________

24. On the figure below, diagram the stages of infection and disease (symptom intensity vs. time). Label each axis, exposure to the microbe, the incubation period, prodromal stage, height of infection, and convalescence.

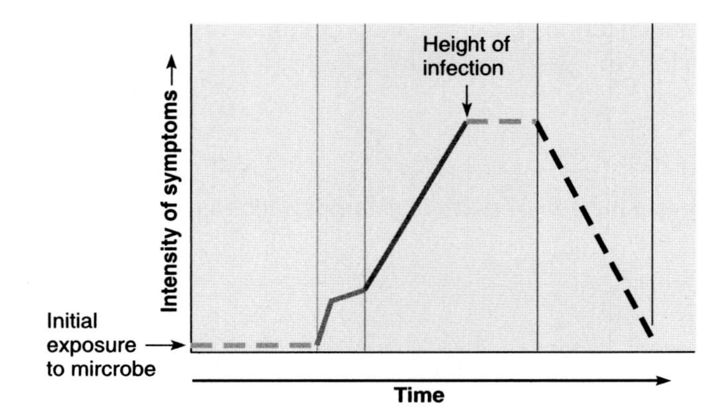

25. A focal infection has characteristics of both localized and systemic infections. How is this the case? (Define localized, systemic and focal infections.)

26. How are a mixed infection and a primary–secondary infection similar?

How are they different?

27. Is the common cold an acute or chronic disease? Why?

28. Differentiate between the signs and symptoms of an illness.

Is a fever a sign or a symptom?

Is a headache a sign or a symptom?

What is a syndrome?

29. Describe four separate portals of exit and the pathogens that use them.

a. ______________________________

b. ______________________________

c. ______________________________

d. ______________________________

30. What are latent infections? Give two examples of latent infections.

a. ______________________________

b. ______________________________

31. What are the sequelae of Strep throat and Lyme disease?

32. Differentiate between a reservoir and a source of infection.

33. What is the difference between asymptomatic carriers and passive carriers?

34. Houseflies and roaches can transmit disease because they are ______________ vectors.

35. The transmission of malaria by mosquito bite is an example of a ____________ vector.

36. Are humans the natural host of rabies and West Nile viruses? Explain.

37. How may an infectious disease be spread from person to person?

How do direct and indirect contact methods of transmission differ?

Give examples of each.

38. What is a fomite?

39. Is tetanus a communicable infectious disease? Explain.

40. Differentiate between aerosol spread and droplet nuclei spread of respiratory pathogens.

Which spread the hardier pathogens?

41. Where are nosocomial infections acquired?

42. What are universal precautions and why are they considered universal?

43. Which branch of microbiology studies the effects of disease in a population?

44. What are reportable diseases?

45. Distinguish between the prevalence and incidence of a disease?

    If there are 20 cases in a population of 100, what is the prevalence of the disease?

    If the following week there are 10 more new cases, what is the incidence?

46. Differentiate between morbidity and mortality.

    Morbidity: ________________________

    Mortality: _________________________

47. If a disease is endemic in an area, what does that indicate?

48. Differentiate between an epidemic and pandemic.

49. What is the "iceberg effect" when applied to epidemiology?

Organizing Your Knowledge
Table 1. Distribution of Flora

Area of the Body	Normal Flora (Yes/No)	Internal or External?
Skin	a.	External
Heart	No	h.
Upper respiratory tract	b.	External
Bones	No	i.
Mouth	c.	External
Muscles	No	j.
Vagina	d.	External
Ovaries	No	k.
Urine (in bladder)	e.	Internal
Lungs	No	l.
Gastrointestinal tract	f.	External
Saliva (in salivary glands)	No	m.
External eye	g.	External
Blood	No	n.

Table 2. Disease Transmission Overview

Entry Point	Pathogen	Disease	Exit
a.	f.	Cholera	k.
b.	*S. pyogenes*	i.	l.
c.	g.	AIDS	m.
d.	*M. tuberculosis*	j.	n.
e.	h.	Tetanus	o.

Self-Test: Vocabulary

1. ______ inanimate object that harbors and transmits disease
2. ______ toxins produced by bacteria that harm white blood cells
3. ______ disease state caused by the ingestion of toxins rather than infection
4. ______ cell and tissue death
5. ______ toxins that burst red blood cells
6. ______ time between contact with infectious agent and appearance of symptoms
7. ______ condition where viruses are found in a patient's blood
8. ______ carrier that transmits disease-causing organisms without being infected
9. ______ substances that are secreted by bacteria to promote invasion
10. ______ animal disease that is naturally transmissible to humans
11. ______ infection acquired in the hospital
12. ______ study of disease in populations
13. ______ number of new cases in a certain period of time
14. ______ a disease that is commonly present in a given area is ____
15. ______ minimum number of organisms needed to cause disease
16. ______ process by which microbes bind to host tissues
17. ______ infection caused by growth of normal flora in normally sterile sites
18. ______ germ-free
19. ______ white blood cells that engulf and destroy bacteria
20. ______ microbes capable of causing disease in healthy individuals

A. adhesion
B. endemic
C. endogenous
D. epidemic
E. epidemiology
F. exoenzymes
G. fomite
H. gnotobiotic
I. hemolysins
J. incidence
K. incubation
L. infectious dose
M. intoxication
N. leukocidins
O. necrosis
P. nosocomial
Q. passive
R. pathogens
S. phagocytes
T. prevalence
U. viremia
V. zoonosis

Self-Test: Multiple Choice

1. A pathogen may be ___.
   a. a fungus
   b. a bacteria
   c. a virus
   d. all of the above
2. Which of the following organisms enters the body via the gastrointestinal tract?
   a. rabies
   b. mycoplasmas
   c. poliovirus
   d. histoplasmas
3. In general, sites of the human body devoid of normal flora are ____.
   a. nonexistent
   b. external body surfaces
   c. internal body surfaces, such as the intestinal tract
   d. internal organs, such as the bladder
4. An indirect method of disease transmission would be by ___.
   a. rabid dog bite
   b. mosquito transmission of malaria
   c. touching a fomite
   d. kissing someone with mononucleosis
5. Two of the heaviest areas of microbial growth are _____.
   a. the intestine and bladder
   b. the intestine and mouth
   c. the skin and bladder
   d. the mouth and bladder
6. Nosocomial infections are acquired ____.
   a. by direct contact with an animal
   b. in hospitals
   c. though sexual activity
   d. by indirect contact with animal feces
7. Which of the following is a sign of an active infectious disease?
   a. fever
   b. fatigue
   c. headache
   d. sore throat
8. Which of the following is NOT an adhesion factor for pathogens?
   a. capsules
   b. flagella
   c. fimbriae
   d. cell walls
9. The transient population of flora ____.
   a. normally grows on humans
   b. is found on deeper layers of human skin and forms a stable population
   c. is acquired by routine contact
   d. is a stable population that causes disease
10. The ____ of a disease is the ratio of the number of new cases to the total number of people in a given population.
    a. incidence
    b. prevalence
    c. morbidity
    d. mortality
11. If a pneumonia-causing microbe causes pneumonia, then worsens to cause septicemia, we would call this a ____ infection.
    a. toxemic
    b. localized
    c. focal
    d. mixed
12. The portal for the greatest number of pathogens is the ___.
    a. respiratory tree
    b. gastrointestinal tract
    c. skin
    d. urogenital (STDs)
13. Endogenous infections are caused by ___.
    a. true pathogens causing disease
    b. fungal infections only
    c. normal flora
    d. viral infections only

14. Gnotobiotic animals _____.

   a. lack normal flora or have well-defined flora
   b. have a well-developed immune system
   c. often have more cavities than normal animals
   d. are less sensitive to gut pathogens like salmonella

15. Intoxications are due to____.

   a. the ingestion of live organisms that grow and produce endotoxin
   b. the ingestion of exotoxins
   c. the growth of toxin-producing bacteria in the blood
   d. the ingestion of live organisms that grow and cause disease

Applications of Chapter 13

1. One goal of epidemiologists is to find the natural reservoir of diseases, particularly those that are prone to sporadic, deadly outbreaks such as Ebola or yellow fever. Why are vector-born diseases particularly difficult to track in a population?

2. There are several reportable diseases in the United States and worldwide. Why are there reportable diseases and why aren't ALL diseases reportable?

Chapter 14 Nonspecific Host Defenses

Building Your Knowledge

1. List the three "lines of defense" a host organism uses to prevent invasion by pathogens.

   a. ______________________________

   b. ______________________________

   c. ______________________________

   Which of these defense mechanisms are innate defenses?

   Which of these are acquired through exposure to a pathogen or vaccine?

2. List several physical barriers to infection in a mammalian body.

3. Humans continuously produce and lose skin cells. How does this aid in the prevention of infection?

   What other barriers to infection does the skin pose?

4. Chemical defenses are chemicals produced by the body to impede the growth of microbes. List three of these defenses, how they impede the growth of microbes, and where they are produced.

Chemical Defense	Antimicrobial Action	Location Produced
a.	c.	Sebaceous glands
b.	Inhibits microbial growth	e.
Lysozyme	d.	f.

5. Genetic defenses may exist between different species (interspecific) or within a single species (intraspecific). Give an example of each.

   a. Interspecific genetic defense: __________________.

   b. Intraspecific genetic defense: __________________.

6. What are the three major tasks the immune system accomplishes for a healthy body?

   a. ______________________________

   b. ______________________________

   c. ______________________________

7. Describe the structure and function of the reticuloendothelial system. Where is the RES in a human body?

   How is the RES connected to the blood system, extracellular fluid, and lymphatic systems?

8. What is hemopoiesis?

   Where does it take place in infants?

   Where does it take place in adults?

9. Granulocytes and agranulocytes are two groups of cells produced. How are they similar?

10. List the three types of granulocytes and the two types of agranulocytes and the function of each cell type.

11. How are monocytes and macrophages related?

12. What is the structure and function of dendritic cells?

13. How are platelets produced and what is their function in the body?

14. Chemotaxis and diapedesis are related processes. How are they similar? How are they different? How are they related?

15. How do primary lymphoid tissues differ from secondary lymphoid tissues?

16. How do macrophages recognize microbes as foreign?

17. What types of molecules do toll-like receptors recognize?

18. What are the three basic functions of an inflammatory response?

a. ______________________________

b. ______________________________

c. ______________________________

19. What are the four cardinal (classic) signs of inflammation, what are the Latin terms for these signs, and what causes these symptoms?

Cardinal Sign	**Latin Term**	**Cause of the Symptom**
a.	Rubor	e.
Pain	c.	Stimulation of nerves
b.	Tumor	Edema
Heat	d.	f.

20. What cells are generally the first to reach the site of inflammation?

Are these cells part of the specific or nonspecific response to invasion?

21. What are the possible causes of a fever?

    How does a fever help fight an infection?

22. Which white blood cells (leukocytes) are phagocytes?

    How do phagocytes kill bacteria?

23. What is the complement cascade? There are two infection-fighting results of this cascade. What are they?

Organizing Your Knowledge

Table 1. Immune Cell Overview

Immune Cell	**Function**	**Phagocyte?**	**Granulocyte?**
Neutrophils	a.	Yes	Yes
B cells	Produces antibodies	d.	No
Monocytes	Phagocytes	Yes	g.
Eosinophils	b.	Minor role	Yes
Mast cells	Produce histamine	e.	Yes
Macrophages	Phagocytes	Yes	h.
T cells	c.	No	No
Dendritic cells	Antigen presentation	f.	No

Self-Test: Vocabulary

ACROSS

1 B cells and T cells
4 leukocytes that make up 55% to 90% of circulating white blood cells
7 production of blood cells
9 an enzyme that breaks down peptidoglycan and is found in tears
11 cicrulating cells that function in allergy (histamine release)
13 white blood cells
16 process by which white blood cells exit circulation
18 receptors on phagocytes that recognize pathogen associated molecules

DOWN

2 chemical mediators released by cells that provide cross-talk between white blood cells.
3 proteins produced by some white blood cells and tissue cells that have antiviral activities.
5 Immunity mediated by B cells
6 system made up of circulating proteins that are activated in a sequential (cascade) fashion to stimulate inflammation
8 bacteria cause a lot of pus formation are __.
10 line of defense that includes physical, chemical and genetic barriers to infection
12 any substance capable of causing a fever
14 site of T cell maturation
15 plasma that lacks clotting factors
17 lymphoid tissues where leukocytes develop and mature

Self-Test: Multiple Choice

1. Granulocytes include _____.
   a. macrophages and neutrophils
   b. monocytes and eosinophils
   c. basophils and lymphocytes
   d. neutrophils and basophils

2. B cells and T cells are ___.
   a. both lymphocytes
   b. both part of cell-mediated immunity
   c. both capable of producing antibody
   d. both granulocytes

3. The nonspecific defenses, such as phagocytes, are part of a body's ___ line of defense against infection.
   a. first
   b. second
   c. third
   d. fourth

4. Cytokines aid in an immune response by ___.
   a. directly killing viruses
   b. activating leukocytes
   c. activating erythrocytes
   d. activating the complement cascade

5. Which of the body's fluid-filled spaces does NOT participate heavily in an immune response?
   a. bloodstream
   b. reticuloendothelial system
   c. lymphatic system
   d. cerebrospinal fluid

6. The lymphatic system includes all of the following EXCEPT ____.
   a. the spleen
   b. lymph nodes
   c. the heart
   d. the thymus

7. Any substance that causes fevers to develop is called a ____.
   a. cytokine
   b. pyogen
   c. pyrogen
   d. interferon

8. Which of the following cells is NOT part of the monocyte line of differentiation?
   a. monocytes
   b. macrophages
   c. platelets
   d. dendritic cells

9. Which of the following is an example of a chemical barrier to infection?
   a. blinking and lacrimation
   b. lysozyme
   c. lack of receptors on humans for distemper
   d. desquamation

10. Leukocytes that are phagocytic are ___.
    a. B cells and neutrophils
    b. platelets and B cells
    c. neutrophils and macrophages
    d. dendritic cells and B cells

11. Which of the following is mismatched?
    a. rubor—redness
    b. tumor—swelling
    c. calor—heat
    d. dolor—pus formation

12. Which of the following is mismatched?
    a. basophils—histamine
    b. neutrophils—specific immunity
    c. macrophages—phagocytes
    d. B cells—humoral immunity

13. A person with a high eosinophil count will likely ____.
    a. have an active helminth infection
    b. have an active histamine response
    c. have an active bacterial infection
    d. develop a pus-filled abscess

14. Human blood consists of ____.
    a. plasma, white blood cells, blue blood cells
    b. fibrin, plasma, white blood cells
    c. white blood cells, red blood cells, plasma
    d. plasma, lysozyme, hematin

15. Immune cells cross blood vessels to enter tissue spaces by ______.

    a. differentiation
    b. phototaxis
    c. complement
    d. diapedesis

Applications of Chapter 14

1. Often inflammation can be treated with steroids that will prevent leukocytes from expressing the adhesion molecules on their surfaces needed for diapedesis. How would this limit an inflammatory response?

2. There are two types of an immune reaction—the adaptive response that develops upon exposure to a specific pathogen and confers immunity to that pathogen and the innate response that recognizes general "danger" signals associated with pathogens. What is the advantage to having two different systems?

Chapter 15 Adaptive, Specific Immunity and Immunization

Building Your Knowledge

1. What is immunocompetence?

2. Which cells are most important in the development and maintenance of specific immunity to antigens?

3. What two features most characterize acquired immunity?

4. Differentiate between active and passive immunization.

   Which is quicker?

   Which is longer lasting?

5. Differentiate between natural and artificial immunity.

6. How does immunotherapy differ from active immunization?

7. Give examples of four separate types of immunization:

Immunization	**Example**
Natural active	
Natural passive	
Artificial active	
Artificial passive	

8. Where do B cells mature?

Where do T cells mature?

9. B cells produce ___________________________ in response to antigen.

10. Receptors are found on all immune system cells. What are four major functions of receptors?

11. Put the following events in order and describe each process.

Event	**Order (First, Second, etc.)**	**Description of Event**
Antigen presentation and clonal selection		
Activation and clonal expansion		
B cell and T cell response		
Lymphocyte development		

12. A specialized group of receptors are MHC antigens. What does MHC stand for?

Draw an MHC I and an MHC II molecule. Which cells express MHC I and which cells express MHC II?

MHC I MHC II

13. What is the clonal selection theory?

Do lymphocytes change their antibodies or other receptors to match a given antigen? Explain.

14. What is immune tolerance and how does it occur?

15. What are immunoglubulins and which cells produce them?

16. Label the following figure, indicating the location of three disulfide bonds, heavy chains, light chains, variable regions, constant regions, the antigen binding site, the Fab area, and the Fc region.

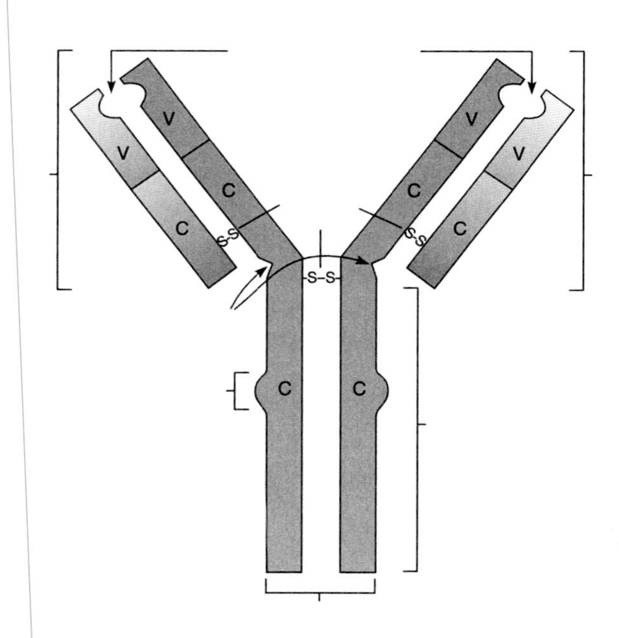

17. How are T cell receptors similar to immunoglobulins?

How are they different?

18. Where do B cells mature?

19. Where do B cells go after maturation?

    Do B cells circulate or "home" to a particular region?

20. Where do T cells mature?

21. In general, what traits make a good antigen rather than a poor antigen?

    What is a hapten?

    How do you produce antibodies against a hapten?

22. Differentiate between autoantigens and alloantigens.

    Which are of concern to transplant specialists?

23. What are superantigens and which group of lymphocytes do they directly stimulate?

24. How do most antigens enter the body?

    Where are they gathered up and concentrated after entry?

25. What are antigen-presenting cells (APCs) and how do they present antigen to the body?

    Which T cells do APCs present antigen to?

26. If an APC presents an antigen to a T-helper cell, how does the APC activate the T cell? How does the T cell then activate a B cell?

27. Do B cells present antigens to T cells on their antibodies?

28. What signals do B cells require to activate?

    Where do these signals come from?

29. What happens to a B cell after activation?

30. What may antibodies do to eliminate a pathogen (list four specific things)?

    a. ______________________________

    b. ______________________________

    c. ______________________________

    d. ______________________________

31. List and differentiate between the five different classes of antibody.

Antibody Class	**Description**
IgG	c.
a.	Pentamer
IgD	d.
b.	Dimer
IgE	e.

f. Which is most prevalent in the blood?

g. Which is found on mucous membranes?

h. Which serves as a B cell receptor?

i. Which causes allergy?

j. Which is produced first, upon exposure to an antigen?

32. Draw an antibody titer graph, with the primary and secondary response. Label the IgM curve, the IgG curve, and the latent period.

33. A secondary immune response is stronger, longer, and quicker than a primary response. Why?

34. There are many different types of T cells. List three and their function.

T cell subset	CD4 or CD8	Function
T_{H1}	a.	T cell help for other T cells
T_{H2}	CD4	c.
T_C	b.	Cytotoxic T cells

35. What is ISG therapy?

How is it administered?

Who is it administered to?

How long does the protection last?

36. How does SIG therapy differ from ISG therapy?

37. What are six characteristics of an ideal vaccine?

a. ______________________________

b. ______________________________

c. ______________________________

d. ______________________________

e. ______________________________

f. ______________________________

38. Give an example of a killed or inactivated vaccine.

How are such vaccines prepared?

39. Give an example of an attenuated vaccine.

What are three advantages of live vaccines?

What are three disadvantages of live vaccines?

40. Which type of vaccines are toxoids (attenuated or inactivated) ?

How are toxoids made?

What do they protect against?

41. List three diseases that for which there is no reliable vaccine.

a. ______________________________

b. ______________________________

c. ______________________________

42. What are "Trojan horse" vaccines and how are they constructed?

What carriers have been used?

What vaccines have been developed and used experimentally?

43. How are DNA vaccines different from attenuated or inactivated vaccines?

44. What is herd immunity?

What does it prevent?

Organizing Your Knowledge

Table 1. Comparison of MHC Molecules

MHC Class	Found on ___ Cells	Presents Antigen to:	Gets Antigen from:
MHC I	a.	c.	e.
MHC II	b.	d.	f.

Table 2. Place an X in all of the cells that match the listed characteristics.

Trait	T_{H1}	T_C	T_{H2}
a. Recognizes MHC II			
b. Has CD4			
c. Has CD8			
d. Recognizes MHC I			
e. Regulates immune reactions			
f. Produces antibody			
g. Provides B cell and T cell help			
h. Produces perforins			
i. Destroys virally infected cells			
j. Reduced in HIV/AIDS patients			

Table 3. Place an X in all of the cells that match the listed characteristics.

Attribute	T Cell Receptor	Antibody	Both
a. Antigen-binding sites			
b. Variable and constant regions			
c. Light or heavy chains			
d. Formed by genetic modifications			
e. Secreted			
f. Binds MHC and antigen			
g. Found on B cells			
h. Recognizes free antigen			
i. Recognizes antigen bound to MHC			
j. Disulfide bonds			

Self-Test: Vocabulary

1. ______ removal of harmful (self-reactive) lymphocytes established ___
2. ______ toxin produced by *S. aureus* that over-stimulates T cells is a ____
3. ______ T cells that provide B cell help are ____
4. ______ immunity that is B and T cell-mediated and confers long-term protection
5. ______ molecules that activate B and/or T cells
6. ______ and _______. two features that characterize adaptive immunity
7. ______ immunity conferred by getting an MMR shot
8. ______ immunity given to nursing infant
9. ______ MHC molecules found on all nucleated cells
10. ______ site of T cell maturation
11. ______ region of an antibody molecule that binds to an antigen
12. ______ small molecules that require a carrier to stimulate an immune response
13. ______ antibodies that neutralize toxins
14. ______ coating bacteria with antibodies, making them better recognized by phagocytes
15. ______ antibody class that is secreted and found on mucosal surfaces
16. ______ antibody class that binds to mast cells and contributes to allergies
17. ______ proteins produced by cytotoxic T cells that "punch holes" in target cells
18. ______ process that lessens the virulence of a pathogen so it can be used as a vaccine
19. ______ collective immunity of a population
20. ______ a compound that enhances the immunogenic properties of an antigen (vaccines)

A. acquired
B. active artificial
C. active natural
D. adjuvant
E. antigens
F. antitoxins
G. attenuation
H. CD4+
I. CD8+
J. Fab
K. Fc
L. hapten
M. herd immunity
N. IgA
O. IgE
P. IgG
Q. memory
R. MHC I
S. MHC II
T. natural passive
U. opsonization
V. perforins
W. specificity
X. superantigen
Y. thymus
Z. tolerance

Self-Test: Multiple Choice

1. Which regions of an immunoglobulin bind to antigen?
    a. variable regions of heavy chains, constant regions of light chains
    b. variable regions of light chains, constant regions of heavy chains
    c. variable regions of both chains
    d. constant regions of both chains

2. Which of the following make good antigens?
    a. starch, with it's repetitive sugars
    b. tetanus toxin, since it's an exotoxin
    c. hemoglobin, with it's repetitive amino acids
    d. oils, with the gycerol and fatty acids

3. B cells and T cells can make up the ___ immune response and are types of ____.
    a. specific: lymphocytes
    b. nonspecific: neutrophils
    c. specific: macrophages
    d. non-specific: eosinophils

4. Antitoxins are used to ____.
    a. actively immunize patients
    b. naturally immunize infants
    c. passively immunize patients
    d. kill cancer cells

5. According to the clonal selection theory, ________.
    a. lymphocytes change to match antigens
    b. naive lymphocytes are a diverse population
    c. a lymphocyte can recognize many different epitopes
    d. macrophages differentiate into T cells

6. Which of the following is NOT an antibody function?
    a. presenting antigen to T helper cells
    b. opsonization
    c. neutralizing toxins
    d. fixing complement

7. If a dendritic cell presents antigen on MHC I, a ____ will _____.
    a. helper T cell: boost a B cell antibody response
    b. cytotoxic T cell: kill nearby B cells
    c. helper T cells: boost the dendritic cells' antibody response
    d. cytotoxic T cell: kill the dendritic cell

8. Which of the following is NOT a B cell activation response?
    a. differentiate to plasma cells
    b. clonal expansion
    c. producing antibody
    d. kill the antigen presenting cell

9. Vaccination strategies that are currently being tested include ____.
    a. DNA vaccines
    b. subunit vaccines
    c. Trojan horse vaccines
    d. all of the above

10. Attenuated vaccines are ____.
    a. live vaccines
    b. heat-killed organisms
    c. bits of immunogenic DNA
    d. made up of antigenic subunits of a pathogenic organism

11. An anamnestic response ________.
    a. lasts longer than a primary response
    b. is stronger than a secondary response
    c. produces only monoclonal antibodies
    d. is primary IgM

12. If a B cell encounters an antigen and presents it to a T helper cell, the antigen will be presented on a ____ molecule.
    a. MHC III
    b. antibody
    c. MHC II
    d. MHC I

13. B cells are __ while T cells are ____.
    a. circulating: located in the thymus
    b. located in the bone marrow only: circulating
    c. circulating: noncirculating
    d. noncirculating: circulating

14. Immunoglobulins are also called ___ and are produced by ___.
    a. lymphokines: T cells
    b. cytokines: B cells
    c. antigens: macrophages
    d. antibodies: B cells

15. If a person had high levels of IgE circulating in his or her bloodstream, this person is likely ___.
    a. fighting a viral infection
    b. in the middle of an allergic response
    c. in the middle of a primary antigen response
    d. in the middle of a secondary antigen response

Applications of Chapter 15

1. If you are bitten by a rattlesnake, would you want the emergency room doctors to actively or passively immunize you? Why?

2. A macrophage is infected with a virus and presents the viral proteins to a T cell. How are these proteins presented, which T cells are activated by this cell, and what is the fate of the macrophage once the T cell has been activated?

Chapter 16 Disorders in Immunity

Building Your Knowledge

1. In what two generalized ways can the immune system malfunction and cause disease?

   a. ____

   b. ____

2. Label the figure below, indicating which disorders are a loss of immunity, which are B cell mediated, which are T cell mediated, and label Type I–Type IV hypersensitivities.

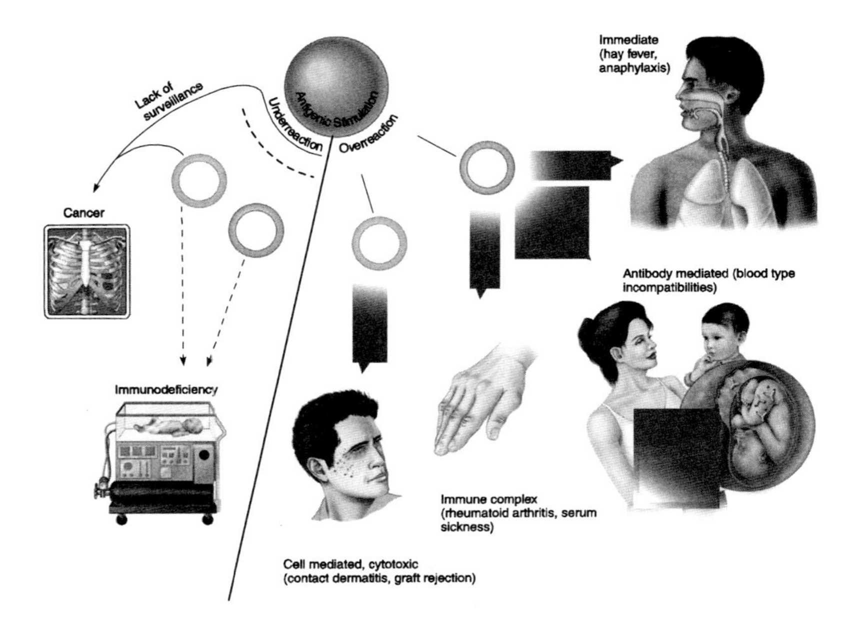

3. How do allergies differ from hypersensitivities? How are they similar?

4. Why is it easy to mistake an allergic reaction for an infection?

5. How does atopy differ from anaphylaxis?

6. How can a child inherit an allergy to pollen from a parent who is allergic to dust mites?

7. Allergy sufferers have a genetic predisposition for _____.

8. How can allergic individuals not have symptoms of allergy on their first exposure to an allergen, such as bee venom, but nearly die from anaphylaxis on their second exposure?

9. List and give examples of four separate portals of entry and allergens.

Entry Portal	**Allergen**
Inhalants	c.
a.	Food, drugs (penicillin)
Injectants	d.
b.	Rubber, solvents, dyes, heavy metals

10. Which antibody class is associated with type I allergies?

Which cells do these antibodies bind to?

11. What compounds are released as a mast cell degranulates?

12. What happens to the human body after mast cell degranulation ?

13. How are asthma, eczema, and food and drug allergies similar?

What are the symptoms of each allergy?

Asthma: ________________________

Eczema: ________________________

Food allergies: ___________________

14. What is the difference between systemic and cutaneous anaphylaxis?

    Which can kill in 15 minutes?

15. How can allergies be diagnosed?

    What molecules or cells are indicative of an active immune response?

16. What three methods are there to treat or prevent allergy attacks?

    a. ____

    b. ____

    c. ____

17. How does desensitization to allergens prevent allergy attacks?

18. How do type II hypersensitivities differ from type I hypersensitivities?

    Which molecules mediate type II hypersensitivities?

19. What are the ABO antigens?

    If a person has an allele for type O and one for type A, what will that person's blood type be?

    Which blood type is the universal donor? Why?

    Which blood type is the universal recipient? Why?

20. What would happen if type B blood was given to someone with type A blood?

21. How is a transfusion reaction treated?

22. What do Rh positive and Rh negative mean?

    How are Rh negative females commonly sensitized to Rh positive antigens?

    How does RhoGAM prevent hemolytic disease of the newborn?

23. What are type III hypersensitivities caused by? How are type II and type III sensitivities similar and how are they different?

24. What are the symptoms of serum sickness?

    What are these symptoms caused by?

25. What is the Arthus reaction?

    How are serum sickness and the Arthus reaction similar?

26. Which cells mediate type IV hypersensitivities?

    Are these immediate or delayed reactions?

27. Which class of T helper cells cause the tuberculin reaction?

    What does a positive tuberculin test indicate?

28. What immune cells and molecules (markers) are responsible for transplant rejection?

29. What cells are transplanted to cause GVHD?

    How is GVHD similar to rejection of foreign tissue?

30. Where may donor organs and tissues come from?

    Which types are more successful?

31. Which type(s) of hypersensitivity are responsible for most autoimmune reactions?

32. Are males or females more commonly diagnosed with autoimmune disease?

33. Describe the four theories of autoimmune reaction development.

34. How are SLE and rheumatoid arthritis similar?

35. Name and briefly describe an endocrine and a neuromuscular autoimmune disease.

36. Differentiate between primary and secondary immunodeficiency diseases.

    Primary: ___________________________

    Secondary: _________________________

37. What diseases are associated with a lack of, or deficiency in, antibody production?

    What infections are these patients prone to get?

38. What is DiGeorge syndrome and what problems are associated with it?

39. What defects may cause SCID?

40. What four general ways can a person acquire an immunodeficiency?

    a. __________________________

    b. __________________________

    c. __________________________

    d. __________________________

41. What exactly is cancer?

    What is the difference between a benign and a malignant tumor?

    What is metastasis?

42. Why are AIDS patients more prone to certain types of cancer?

Organizing Your Knowledge

Table 1

Immunopathology	Overactivity or Underactivity?
Allergies	a.
Graft rejection	b.
Immunodeficiency	c.
Autoimmunity	d.

Table 2. Overview of Hypersensitivities

Hypersensitivity Reaction	Cells or Antibodies Involved	Type of Hypersensitivity
Asthma	a.	I
Rheumatoid arthritis	Antibodies and cells	f.
Contact dermatitis	b	IV
Hay fever	Antibodies and cells	g.
Graft rejection	c.	IV
Serum sickness	Antibodies	h.
Anaphylaxis	d.	I
Eczema	Mast cells	i.
Arthus reaction	e.	III

Table 3. Transplant Overview

Graft Type	Donor	Example
Xenograft	a.	e.
Isograft	b.	f.
Allograft	c.	g.
Autograft	d.	h.

Self-Test: Vocabulary

ACROSS

5 chronic local allergy
7 treatment of allergy by administering shots of specific allergens
11 test based on a DTH response due to exposure to tuberculosis
15 antibody responsible for allergic responses
17 immunodeficiencies that are present from birth
18 graft vs host disease
19 state where immune response is not adequate
20 tissue grafted from another species

DOWN

1 localized type III hypersensitivity
2 respiratory disease with impaired breathing due to bronchoconstriction
3 preventative treatment for erythroblastosis fetalis
4 delayed reaction that is an immune overreaction to a stimulus
6 a person with type O blood is a universal _____
8 disease associated with over or under activity of immune system
9 dose of allergen that will cause symptoms to appear
10 antigens that stimulate an allergic response in some individuals
12 release of substances from mast cells
13 systemic (often fatal) reaction affecting respiratory and circulatory systems
14 atopic dermatitis
16 fastest acting allergic mediator released from mast cells

Self-Test: Multiple Choice

1. A person with hay fever may pass on a mold allergy to his or her children.
    a. True
    b. False

2. Which of the following is NOT typically an inhaled allergen?
    a. pollen
    b. animal hair
    c. antibiotics
    d. mold spores

3. Immediate hypersensitivities are caused by ____ reactions.
    a. IgE
    b. neutrophil
    c. macrophage
    d. IgM

4. Which of the following is NOT an autoimmune disease?
    a. rheumatoid arthritis
    b. DiGeorge syndrome
    c. diabetes mellitus
    d. multiple sclerosis

5. Antigens that arise from self tissue and cause a hypersensitivity are ___ and cause ____ reactions.
    a. exogenous: IgE
    b. endogenous: IgE
    c. exogenous: autoimmune
    d. endogenous: autoimmune

6. A person with type A blood will have antibodies against ____.
    a. type B blood
    b. type O blood
    c. type A blood
    d. none of the above—type A is the universal donor

7. The sequestered antigen theory explains the development of ___.
    a. cancer
    b. acquired immunodeficiency
    c. autoimmune disease
    d. allergic hypersensitivities

8. Which of the following are type I hypersensitivities?
    a. food allergies
    b. asthma
    c. eczema
    d. all of the above

9. It is often difficult to distinguish an allergic response from an active infection because ____.
    a. both are contagious
    b. both are inflammatory responses
    c. both cause cancer
    d. both are mediated by red blood cells

10. A transplant from one species to another, such as the use of pig heart valves in humans, is called a(n) ___.
    a. allograft
    b. xenograft
    c. isograft
    d. autograft

11. RhoGAM is used to treat ___ mothers of ___ children, to prevent damage to future offspring.
    a. Rh+ , Rh-
    b. Rh-, Rh+
    c. AB+, O-
    d. A, B

12. Which of the following is NOT a mediator of type I hypersensitivity?
    a. antigen–antibody complexes
    b. IgE
    c. histamine
    d. bradykinin

13. Serum sickness and the Arthus reaction are both ___.
    a. systemic diseases
    b. type I hypersensitivities
    c. autoimmune reactions
    d. type III hypersensitivities

14. A primary immunodeficiency ___.
    a. is acquired through repeated exposure to allergens
    b. is present from birth and may lead to anaphylaxis
    c. is present from birth and may lead to opportunistic infections
    d. is acquired through exposure to immunosuppressive drugs, such as steroids

15. How may allergies be treated or prevented?
    a. hypersensitization
    b. taking histamines
    c. desensitization therapy
    d. tissue transplantation

Applications of Chapter 16

1. Children who have chronic, severe asthma often have either eczema and/or allergies as well. Why do you think this is the case? Given this fact, could desensitization therapy benefit eczema or asthma suffers? Why or why not?

2. An Rh- mother did not know her mate was Rh+ and gave birth to an Rh+ daughter. Why was this child not suffering from hemolytic disease of the newborn? Will her next child be at risk? Can she be given RhoGAM to prevent future problems? Why or why not?

3. A small boy grabbed a branch and started chasing his classmates with it. They ran and said he was going to be sorry for touching the poison ivy. He laughed and said, "If this was poison ivy, could I do this? . . . and rubbed it all over his face." Six hours later, his hands and face were still clear of itching, inflammation, or rash—was he correct or were his classmates right to run? Explain.

Chapter 17 Diagnosing Infections

Building Your Knowledge

1. What are the three categories of tests microbiologists use to identify bacteria?

    a. ______________________

    b. ______________________

    c. ______________________

2. What traits can be identified by microscopic morphology?

    What traits are determined by colony morphology?

3. How do biochemical traits aid in bacterial identification?

4. Why are genotypic methods of bacterial classification at times preferred to phenotypic methods?

5. Why are immunologic methods often used to diagnose viral infections?

6. What is the first step in identification of pathogenic organisms in a clinical setting?

7. How are samples collected from a human body? Describe the collection procedure for several different sites.

8. How are most samples transported and stored prior to analysis?

9. Are results from direct testing or culture-based tests known first?

10. Differentiate between presumptive and confirmatory data.

11. How long does it take to completely analyze a sample?

12. Do serological tests require the culture and isolation of a microbe?

13. What are some organisms that can be tested for by direct antigen testing?

    What are the advantages to this type of testing?

    Can you have a negative direct test but a culture that is positive for the organism? Why or why not?

14. Why are isolation media used for fecal samples?

15. What does a positive biochemical test generally indicate?

    How are biochemical profiles determined in clinical labs?

16. What is phage typing?

17. Would you rather have a patient with a sputum culture with 10 *M. tuberculosis* in it or a patient with 200 *E. coli* in his urine culture? Explain.

18. Describe four separate ways genotypic methods are used to identify bacteria.

19. What is serologic testing?

20. Which bodily fluids or tissues can be used to test the immunologic status of an individual?

21. Can you test for unknown bacteria with patient-derived antibodies? Why or why not?

22. Differentiate between sensitivity and specificity.

    Can a test be sensitive but not specific? Give an example.

    Can a test be specific but not sensitive? Give an example.

23. How do you determine a titer?

    What does a high titer indicate?

24. How do agglutination assays differ from precipitation assays?

25. List three separate agglutination tests.

26. What is an Ouchterlony test and what is it used for?

27. What is the advantage of a Western blot over an ELISA for testing for HIV?

28. Diagram the procedure followed in a complement fixation test.

Does hemolysis indicate the patient serum had antibodies in it?

29. Describe the procedures used to test for the presence of :

Salmonella: ________________________________

Syphilis: ____________________________________

Pneumococcus: _______________________________

30. Diagram the processes of direct and indirect immunofluorescence testing.

31. Diagram the process followed for indirect and sandwich ELISA tests.

Which test is used to determine the antigen level in a sample?

How are ELISA tests used in clinical labs? Give several examples.

32. How can we differentiate between T cells and B cells in the lab? Describe two methods.

33. What is the tuberculin test and what does it detect?

Organizing Your Knowledge

Table 1. Sampling Techniques

Sampling Site	Method of Collection
Blood	a.
Nasopharynx	b.
Saliva	c.
Skin	d.
Spinal column	e.
Throat	f.
Urine	g.
Vagina	h.

Table 2. Precipitation vs. Agglutination

Type of Test	Size of Antigen	Solubility of Antigen	Example of Use
Precipitation	a.	c.	e.
Agglutination	b.	d.	f.

Self-Test: Vocabulary

1. ______ methods that assess physical characteristics of microbes (microscopic and macroscopic morphology, biochemical properties, and chemical composition).
2. ______ morphological traits that include colony shape, size, color, and speed of growth
3. ______ media that maintains microbes but does not allow for microbial growth
4. ______ tests using fluorescent antibodies directly on a sample and examining the sample (used for syphilis tests)
5. ______ test used to confirm a positive HIV ELISA
6. ______ the ability of a test to detect small amounts of a pathogen or antibody
7. ______ small, labeled fragments of DNA that hybridize with a test sample.
8. ______ clumping of cells in an antigen–antibody reaction
9. ______ branch of immunology involved within vitro testing of serum
10. ______ enzyme-linked immunosorbent assay

A. agglutination
B. direct fluorescence antigen
C. ELISA
D. genotypic
E. isolation media
F. lysin
G. macroscopic
H. microscopic
I. phenotypic
J. precipitation
K. probes
L. sensitivity
M. serology
N. specificity
O. transport media
P. Western

Self-Test: Multiple Choice

1. A double-diffusion or Ouchterlony test is a type of ____ test.
    a. agglutination
    b. vaccination
    c. ELISA
    d. precipitation

2. The complement fixation test detects the presence of ____ in serum.
    a. toxins
    b. antigen
    c. antibody
    d. bacterial cells

3. If a test can detect very small amounts of an antigen, such as tetanus toxin, but also detects botulism toxin, we can say that it has__.
    a. sensitivity without specificity
    b. titers without sensitivity
    c. specificity without sensitivity
    d. specificity without titers

4. Which type of white blood cell readily forms rosettes when mixed with sheep red blood cells?
    a. neutrophils
    b. T cells
    c. B cells
    d. macrophages

5. Which of the following items would NOT be the target of an agglutination test?
   a. red blood cells
   b. bacterial cells
   c. exotoxins
   d. latex beads

6. True or false. Indirect ELISAs are commonly used to determine the level of antigen in a sample.
   a. True
   b. False

7. Radioimmunoassays (RIA) are preferred over an older test, such as complement fixation because _________.
   a. there is a shortage of complement in most labs
   b. RIA are more sensitive than complement tests
   c. complement tests are more specific than RIAs
   d. RIA detect the DNA that codes for any antigen

8. A person with an antibody titer of 1:128 has ____ than a person with a titer of 1:16.
   a. more antigen in his or her blood
   b. less antigen in his or her blood
   c. less antibody in his or her blood
   d. more antibody in his or her blood

9. After a person tests positive for HIV by ELISA, what is the next step to a confirmatory diagnosis?
   a. a direct ELISA rather than an indirect ELISA
   b. an Ouchterlony test
   c. a Western blot
   d. a complement fixation test

10. Which of the following traits are visible by microscopic morphology?
   a. presence of a capsule
   b. colony pigmentation
   c. catalase positive
   d. ferments glucose

11. True or false. Most bacteria are culturable.
   a. True
   b. False

12. Which of the following samples is normally sterile in healthy humans?
   a. saliva
   b. nasal sputum
   c. cerebrospinal fluid
   d. feces

13. Which of the following pathogens can be tested for by a rapid testing method?
   a. *Streptococcus pyogenes*
   b. HIV
   c. measles
   d. all of the above

14. Phage typing is commonly used to characterize samples of ____.
   a. polio
   b. salmonella
   c. mumps
   d. streptococcus

15. Serological testing is used to identify ____ .
   a. antigens a person has in his or her blood
   b. antibodies a person has in his or her blood
   c. viral titer in a person's blood
   d. bacterial burden in a person's blood

Applications of Chapter 17

1. Stool cultures are very rarely done in clinical settings. Why is this the case?

2. A person can test positive for HIV by ELISA and confirmed by Western blot, but not have detectible virus in his or her blood. Why not?

Chapter 18 The Cocci of Medical Importance

Building Your Knowledge

1. What are the pyogenic cocci? (List the four genera discussed in your book.)

   a. ______________________________

   b. ______________________________

   c. ______________________________

   d. ______________________________

2. What do the staphylococci look like?

   Where are they found?

   Do s*taphylococci* form spores?

   Are they generally flagellated or encapsulated?

   Which staphylococcus is considered the most serious human pathogen from the group?

3. How well does *S. aureus* handle environmental stress?

   Why is this significant?

4. List four separate enzymes produced by *S. aureus* and how these enzymes contribute to the disease process.

5. What specific toxins are associated with *S. aureus* ?

6. How common is staphylococcal colonization?

7. What is MRSA and how is it spread?

8. Differentiate between a furuncle and carbuncle.

9. Systemic staphylococcal infection can take many forms. Describe three.

10. What traits of *S. aureus* make it a leading cause of food poisoning?

11. Are antibodies, cell-mediated immunity, or phagocytes most protective against *S. aureus* infection?

12. Name two coagulase negative staphylococci and the diseases they cause.

    a. ______________________________

    b. ______________________________

13. What test can differentiate staphylococci from streptococci?

14. What is the easiest way to distinguish *S. aureus* from the other staphylococci of medical concern?

15. Differentiate between coagulase and catalase enzymes. What is the function of each?

16. How are staphylococcal infections treated?

    Is penicillin the drug of choice? Why or why not?

17. How may the spread of *S. aureus* be limited?

18. What are the general characteristics of streptococci?

    Do streptococci grow as easily as staphylococci in lab?

19. How are streptococci classified? (Describe two methods.)

20. Differentiate between alpha-hemolysis and beta-hemolysis.

21. What are the streptococci most commonly associated with human disease?

22. How do M proteins and capsules contribute to streptococcal pathogenesis?

23. What toxins and enzymes are produced by streptococci?

24. How does streptococcal infection spread?

    What types of infections are seen most in the summer?

    In winter?

25. Dfferentiate between pyoderma, streptococcal pharyngitis, and scarlet fever.

26. What are the potential long-term complications of streptococcal infection?

27. What antibodies give long-term protection against group A streptococcal infection?

28. Name a group B streptococcus.

    Where are these streptococci found?

    What populations are most at risk for group B streptococcal infection?

29. What are the enterococci?

Where are they found?

Why are they of concern to human health?

30. Why are rapid strep tests so valuable?

31. How are rheumatic fever and glomerulonephritis treated?

32. What are viridans streptococci and what problems do they cause?

33. Why is *S. pneumoniae* also called pneumococcus?

34. Are pathogenic strains of *S. pneumoniae* smooth or rough? Why?

35. How is pneumococcus spread?

Are fomites a primary source of infection? Explain.

36. What three diseases are most commonly associated with pneumococcal infection?

How are pneumococcal infections prevented and treated?

37. Draw and indicate the color of Gram-stained *Neisseria* species.

38. How are *Neisseria* grown in lab? Are they easy or difficult to culture?

39. What species may become infected with *N. gonorrhoeae* ?

    How is the bacterial disease spread?

    What are the symptoms of infection?

    What areas of the body are commonly affected?

40. Differentiate between meningococcus, gonococcus, and pneumococcus.

41. Who is most at risk for meningococcal disease?

    How is meningococcal disease spread?

    What are the symptoms of infection with meningococcus?

    How is meningococcal infection treated?

Organizing Your Knowledge

Table 1

Protein Produced by *S. aureus*	**Action in the Body**	**How This Aids the Bacteria**
Coagulase	Coagulates plasma	c.
Hyaluronidase	a.	Helps invade tissues
Penicillinase	None directly, inactivates penicillin	d.
Leukocidin	b.	Immune evasion
TSST	Superantigen	e.

Table 2

Diseases	Areas Affected	Symptoms
Folliculitis	Skin	c.
Osteomyelitis	a.	Fever. chills, pain, muscle spasms
Bacteremia	Blood—systemic	d.
Food intoxication	b.	Diarrhea , vomiting, cramps
Scalded skin syndrome	Skin	e.

Table 3

Organism	*S. aureus*	*S. pyogenes*	*S. pneumoniae*	*N. gonorrhoeae*	*N. meningitidis*
Gram stain	a.	Positive	e.	Negative	i.
Disease caused	"Staph" infections	c.	Pneumonia, meningitis, otitis media	g.	Meningitis
Diagnostic test	b.	Rapid tests	f.	Gram-negative diplococci in exudates	j.
Treatment	Antibiotics—after resistance profile developed	d.	Antibiotics—after resistance profile-developed vaccine	h.	Penicillin G, vaccine

Self-Test: Vocabulary

1. ______ delayed inflammation of the joints, heart, and subcutaneous tissues
2. ______ toxin that causes scarlet fever
3. ______ microbe with a wide range of growth conditions that grows in grapelike clusters
4. ______ toxins that lyse red blood cells
5. ______ enzyme that digests hyaluronic acid
6. ______ enzyme that clots blood plasma
7. ______ alpha-hemolytic streptococci that may cause subacute endocarditis
8. ______ large, deep lesion (a cluster of furuncles)
9. ______ stimulation of pus formation
10. ______ infection of bone tissue
11. ______ inflammation of the heart
12. ______ fastidious bacteria that grow in long gram-positive chains
13. ______ protein that helps streptococcus evade phagocytosis and adhere
14. ______ ear infections common in infants and toddlers
15. ______ conjugate vaccine agains streptococcus

A. carbuncle
B. cephalosporin
C. coagulase
D. endocarditis
E. gonorrhea
F. hemolysins
G. hyaluronidase
H. osteomyelitis
I. otitis media
J. Prevnar
K. protein G
L. protein M
M. pyogenic
N. pyrogenic
O. rheumatic fever
P. S. aureus
Q. streptococci
R. viridans

Self-Test: Multiple Choice

1. Group B streptococcal infections often cause serious problems in ___.
   a. college students
   b. adult females
   c. newborns
   d. sexually active males

2. Gonococcus is a ____ that causes ____.
   a. gram-negative coccus: ear infections
   b. gram-positive coccus: pneumonia
   c. gram-negative coccus: gonorrhea
   d. gram-positive coccus: gonorrhea

3. The most powerful defense humans have against *S. aureus* is ___.
   a. antibody production by B cells
   b. cytotoxic T cell killing
   c. neutrophils and macrophages
   d. mast cells and eosinophils

4. Gonococcus, meningococcus, and pneumococcus are all ___.
   a. causative agents of strep throat
   b. gram-negative
   c. gram-positive
   d. cocci

5. Gonococcal infections can cause blindness in infants.
    a. True
    b. False

6. Penicillin is the drug of choice for treating *S. aureus*.
    a. True
    b. False

7. Smooth strains of *S. pneumonia* _____ and are ___.
    a. have pili: virulent
    b. lack a capule: avirulent
    c. have a capsule: virulent
    d. lack pili: avirulent

8. Which of the following characteristics is typical of *S. aureus*?
    a. low salt tolerance
    b. coagulase negative
    c. spore-forming
    d. toxin-producing

9. Pneumococcus commonly causes ____ and pneumonia.
    a. endocarditis
    b. otitis media
    c. dental caries
    d. kidney infections

10. The causative agent of "strep throat" is a ___.
    a. streptococcus
    b. staphylococcus
    c. meningococcus
    d. streptobacillus

Applications of Chapter 18

1. One day you heard a student in a microbiology class claim they heard "a friend of a friend" got gonorrhea from a toilet seat. Based on your knowledge of *Neisseria* biology, how would you respond to this?

2. Penicillin was once the treatment of choice for pneumonia and gonorrhea. Why is it no longer the first drug prescribed for these conditions?

Chapter 19 The Gram-Positive Bacilli of Medical Importance

Building Your Knowledge

1. What is an endospore?

   From the bacterial standpoint, what is the advantage to producing endospores?

   Do ALL gram-positive bacilli form endospores?

2. List the three gram-positive genera that are endospore formers.

   a. ________________________________

   b. ________________________________

   c. ________________________________

3. What are the symptoms of anthrax?

   In which regions, both worldwide and in the United States, is anthrax commonly found?

   What microbe causes anthrax and where does that microbe commonly live?

   How do people in the United States commonly (naturally) acquire anthrax?

   In what forms may an anthrax infection be found? Which is the most serious?

   How is anthrax treated and controlled?

4. How can you differentiate between *Bacillus* and *Clostridium*?

   What is the causative agent of gas gangrene?

   Where is this bacterium commonly found?

   How is gas gangrene treated and prevented?

Why does hyperbaric oxygen therapy effectively treat *Clostridium*?

Why doesn't immunization work to prevent gas gangrene?

5. What disease is caused by *Clostridium difficile*?

   Who is at risk to develop this disease?

6. What bacterial species causes tetanus?

   Do these bacteria form endospores? Why is this important?

   Are scrapes or puncture wounds more likely to become infected with the tetanus bacilli? Why?

   What are the symptoms of tetanus?

   What is the fatality rate for tetanus?

   How is tetanus prevented and treated?

7. What microbe is the causative agent of botulism?

   Is botulism an infection or an intoxication? Explain.

   How is infant botulism contracted?

   How is botulism prevented and treated?

8. How can clinical labs differentiate between the different clostridial species?

9. What are the pathogenic regular-genera of non-spore-forming gram-positive bacilli?

   a. ______________________________

   b. ______________________________

10. What is listeriosis and how is it contracted?

    What microbe causes listeriosis?

    Who is most at risk to develop complications from listeriosis?

    How is listeriosis spread from one cell to another?

11. What disease does *Erysipelothrix rhusiopathiae* cause?

    Who is most at risk to contract the disease?

    Which species generally contract erysipeloid?

    How is the disease treated?

12. What are the genera of the pathogenic irregular non-spore-forming gram-positive bacilli of medical concern?

    a. ______________________________

    b. ______________________________

    c. ______________________________

13. What is the causative agent of diphtheria?

    How is diphtheria transmitted?

    What is diphtherotoxin and how does it contribute to the disease process?

    How is diphtheria diagnosed and treated?

14. What bacterial species is most associated with the development of acne?

15. What are the distinguishing characteristics of the mycobacteria?

    Do mycobacteria produce spores?

    Where do most mycobacteria live in the environment?

    Why are mycobacteria acid-fast?

16. What is the causative agent of tuberculosis?

    Is tuberculosis an emerging infectious disease or ancient disease? Explain.

    Do all infected individuals become ill?

    What is DOT and why is it important in the treatment of tuberculosis?

17. Where can infection with tuberculosis cause symptoms?

    What is a tubercle and how is it formed?

    How is tuberculosis diagnosed?

18. What is the causative agent of leprosy?

    Where is leprosy endemic?

    What animals are used to study leprosy infection?

19. What are the major differences between tuberculoid and lepromatous leprosy?

20. How is leprosy treated?

21. What diseases are caused by pathogenic actinomycetes?

22. What diseases are caused by pathogenic *Nocardia* ?

Organizing Your Knowledge

Table 1

Organism	Shape	Differential Stain Traits	Spore Former
Bacillus anthracis	a	Gram +	k.
Clostridium perfringens	Bacillus	f.	Yes
Clostridium difficile	b.	Gram +	l.
Clostridium tetani	Bacillus	g.	Yes
Clostridium botulinum	c.	Gram +	m.
Mycobacterium tuberculosis	Irregular, bacillus	h.	No
Mycobacterium leprae	d.	Acid-fast	n.
Listeria monocytogenes	Regular bacillus	i.	No
Erysiplothrix rhusiopathiae	e.	Gram +	o.
Corynebacterium diphtheriae	Irregular, bacillus	j.	No

Table 2. Disease Overview

Organism	**Disease(s) Caused**	**Populations at Risk**	**Treatment**
Bacillus anthracis	Anthrax	d.	Ciprofloxacin, supportive care
Clostridium perfringens	Gas gangrene	Wounded	g.
Clostridium difficile	a.	Postantibiotics	Fluids
Clostridium tetani	Tetanus	e.	Supportive care
Clostridium botulinum	Botulism	Infants	h.
Mycobacterium tuberculosis	b.	Endemic regions	DOT—long-term antibiotic
Mycobacterium leprae	Leprosy	f.	Rifampin
Corynebacterium diphtheriae	c.	Unvaccinated	Azithromycin

Self-Test: Vocabulary

ACROSS

1 respiratory disease that produces a toxin that inhibits protein synthesis

4 infant form of this disease is caused by eating raw honey

6 removal of diseased tissue

10 woolsorter's disease

11 chronic progressive disease of the skin and nerves

12 food-borne disease that is of particular danger to pregnant women

DOWN

2 infection sites seen in the lungs of TB patients

3 disease caused by C. difficile

5 clostridial disease with spastic paralysis

7 environmental stress causes these to form

8 form of anthrax characterized by eschar formation

9 fatty acids that contribute to the acid-fastness of mycobacteria

Self-Test: Multiple Choice

1. *Clostridium perfringens* causes ___.
   a. food intoxication and tetanus
   b. leprosy and gangrene
   c. gangrene and food intoxication
   d. tetanus and leprosy

2. Persons completing a course of broad-spectrum antibiotics are at risk to develop ___.
   a. tetanus
   b. pseudomembranous colitis
   c. tuberculosis
   d. leprosy

3. Gram-positive rods can be separated into __.
   a. endospore formers and non-endospore formers
   b. regular endospore formers and irregular endospore formers
   c. acid-fast spore formers and gram-negative spore formers
   d. cell wall positive and cell wall negative spore formers

4. Treatment for leprosy and tuberculosis lasts for ___.
   a. a few days
   b. several weeks
   c. several months
   d. there is no treatment for either disease

5. *Bacillus cereus* and *Bacillus anthracis* both __.
   a. cause pulmonary anthrax
   b. form endospores
   c. stain as acid-fast rods
   d. have a well-developed capsule

6. *Erysipelothrix rhusiopathiae* causes disease in ___.
   a. poultry and humans
   b. humans and cattle
   c. cattle and pigs
   d. humans and pigs

7. The causative agent of acne is a(n):
   a. *Actinomycete*
   b. *Corynebacterium*
   c. *Mycobacterium*
   d. *Proprionibacterium*

8. The most potent microbial toxin known is produced by ___.
   a. *flesh-eating bacteria*
   b. *Clostridium perfringens*
   c. *Clostridium botulinum*
   d. *Mycobacterium leprae*

9. The causative agent of leprosy ___.
   a. forms endospores.
   b. is closely related to the causative agent of anthrax
   c. is acid-fast
   d. has a regular bacillus shape

10. A positive Mantoux test indicates a person ___.
    a. has been exposed to diphtheria
    b. has tetanus
    c. has been exposed to tuberculosis
    d. has leprosy
    e. has tuberculosis

11. Diphtheria is a very serious disease because ___.
    a. it causes lockjaw
    b. it produces a toxin that stops protein synthesis
    c. it produces a neurotoxin
    d. it converts into a deadly virus once inside a host

12. Which of the following infections is routinely prevented by vaccination?
    a. leprosy
    b. tuberculosis
    c. gas gangrene
    d. tetanus

13. The causative agents of tetanus and botulism are both __.
    a. obligate anaerobes
    b. obligate aerobes
    c. mycobacteria
    d. food-borne pathogens

14. The causative agents of tuberculosis and tetanus both:
    a. are acid-fast
    b. form endospores
    c. are obligate anaerobes
    d. are bacilli
    e. all of the above

15. Listeriosis is a(n) ____.
    a. form of tuberculosis
    b. food-borne infection
    c. type of wound infection
    d. intoxication, not infection

Applications of Chapter 19

1. Is culture a common way to determine whether a patient has tuberculosis? Why is this important?

2. Why is anthrax also called "woolsorter's disease"?

Chapter 20 Gram-Negative Bacilli of Medical Importance

Building Your Knowledge

1. What two criteria are used to separate the pathogenic gram-negative bacilli into different categories?

2. Where are enteric bacteria found?

3. Why is lipopolysaccharide (LPS) dangerous?

   Where is LPS found?

4. Where is *Pseudomonas aeruginosa* normally found in the environment?

   What diseases does it cause?

   Who is most at risk to develop a *Pseudomonas* infection?

5. Who is at risk for *Burkholderia cepacia* infection?

6. What are the common names for brucellosis?

   How are pigs and cows affected by brucellosis?

7. How are humans affected by brucellosis?

   How is brucellosis diagnosed and treated?

8. What microbe causes rabbit fever?

Where is tularemia endemic?

What are the major symptoms and treatment for tularemia?

9. What is the causative agent of whooping cough?

   How does infection lead to the "whoop" sound associated with whooping cough?

   Why is whooping cough on the rise?

   How is whooping cough treated?

10. How did *Legionella* get its name?

    Where is *Legionella* commonly found in the environment?

    How is Legionnaires disease diagnosed and treated?

11. What group of pathogens commonly causes diarrheal disease?

    How many diarrheal infections are estimated to occur each year?

12. What is the difference between coliforms and noncoliforms?

    How can you distinguish an *E. coli* from a *Salmonella*?

13. What is enrichment media and why is it used?

14. Draw a gram-negative enteric rod, labeling the K, O, and H antigens.

15. Why is *E. coli* sometimes considered the predominant bacterial species in the human intestine?

16. How do enterotoxigenic, enteroinvasive, and enteropathogenic *E. coli* differ from one another?

17. What is the single greatest cause of death among babies (worldwide)?

    What is the most common cause of "traveler's diarrhea"?

    Why is it better to use Pepto-Bismol than kaolin to treat traveler's diarrhea?

    What is the most common source of urinary tract infections (UTIs)?

18. What diseases do *Klebsiella, Serratia,* and *Citrobacter* cause?

19. How are *Salmonella* and *Shigella* different from the coliforms?

20. What is the causative agent for typhoid fever?

    How is typhoid fever transmitted from host to host?

    What are the major symptoms of typhoid and how is it treated?

21. How do enteric fevers differ from typhoid fever?

    Which animals are particularly associated with *Salmonella* infection?

22. What disease does *Shigella* cause?

What is the natural host for *Shigella*?

How does shigella infection differ from *Salmonella* infection?

What is the treatment for shigellosis?

23. What are the enteric *Yersinia*?

What disease do they cause?

24. What is the causative agent of plague?

What are the virulence factors associated with plague (list three)?

a. ____________________________

b. ____________________________

c. ____________________________

Where is plague endemic?

25. Are humans endemic reservoirs for plague? Explain.

Differentiate between bubonic and pneumonic plague.

Why is plague also called "the Black Death"?

26. How is plague treated?

What is the survival rate, with treatment?

27. List three species of *Haemophilus* and the diseases they cause.

Haemophilus	**Disease Caused**
H. influenzae	b.
a.	Pinkeye
H. ducreyi	c.

Organizing Your Knowledge

Organism	**Disease Caused**	**Reservoir and Transmission**	**At-Risk Populations**
Bordetella pertussis	Whooping cough	Humans, droplet	l.
Brucella abortus	Malta fever	g.	Farmers, butchers
Burkolderia cepia	a.	Soil, contact exposure	m.
Escherichia coli	Diarrhea	h.	Malnurshied, overcrowded
Francisella tularensis	b.	Rabbits (rodents.	Butchers, hunters
Haemophilus aegyptius	Pinkeye	i.	Children, child care worker
Haemophilus influenzae	Meningitis	Droplets	n.
Klebsiella pneumoniae	c.	Droplets	Chronic lung disease patients
Legionella pneumophila	Pneumonia	j.	Immunocompromised
Pseudomonas aeruginosa	d.	Soil, contact exposure	o.
Salmonella	Typhoid	Humans, fecal–oral	p.
Shigella	e.	Humans, fecal–oral	Overcrowded, poor sanitation
Yersinia enterocolitica	Abdominal pain	k.	Those exposed
Yersinia pestis	f.	Rodents (fleas), bite	q.

Self-Test: Vocabulary

1. _____ LPS is also called____
2. _____ an infection caused by medical treatment
3. _____ soil microbe that is a common cause of burn infections
4. _____ opportunistic respiratory pathogen that causes respiratory distress, especially in cystic fibrosis patients
5. _____ disease caused by *Brucella* species
6. _____ a zoonotic disease found in rodents (especially rabbits) in the Northern Hemisphere.
7. _____ other name for whooping cough
8. _____ normal habitat for *Legionella*
9. _____ gram-negative enteric bacteria that ferment lactose
10. _____ causative agent of typhoid is a species of____
11. _____ plague that is transmitted human-to-human
12. _____ blood-loving bacilli
13. _____ causative agent of pinkeye *H.* ____.

A. *aegyptus*
B. aqueous habitats
C. bubonic
D. *Burkholderia cepia*
E. coliforms
F. *ducreyi*
G. endotoxin
H. *Haemophilus*
I. iatrogenic
J. pertussis
K. pneumonic
L. *Pseudomonas aeruginosa*
M. *Salmonella*
N. *Shigella*
O. soil
P. tularemia
Q. undulant fever

Self-Test: Multiple Choice

1. The DTaP vaccine protects against ____.
   a. diphtheria, polio, tetanus
   b. polio, tuberculosis, diphtheria
   c. diphtheria, tetanus, pertussis
   d. diphtheria, tuberculosis, pertussis

2. Both Legionnaire's pneumonia and Pontiac fever are ___.
   a. respiratory illnesses
   b. caused by enteric pathogens
   c. diarrheal illnesses
   d. caused by *Legionella*

3. The causative agents of both diphtheria and whooping cough are _______.
   a. gram-positive
   b. gram-negative
   c. bacilli
   d. sporeformers

4. Which of the following is NOT an enteric pathogen?
   a. *E. coli*
   b. *Salmonella*
   c. *Shigella*
   d. *Haemophilus*

5. Vaccines to protect against *Pseudomonas* infection have been tried in __ and were __.

    a. pregnant women: unsuccessful
    b. cattle farmers: successful
    c. cystic fibrosis patients: successful
    d. burn victims: unsuccessful

6. *Serratia, Citrobacter,* and *Enterobacter* are all ___.

    a. true pathogens
    b. obligate aerobes
    c. opportunistic coliforms
    d. nonenteric, noncoliforms

7. Which of the following is the causative agent of the plague?

    a. *Yersinia pestis*
    b. *Shigella flexni*
    c. *Klebsiella pestis*
    d. *Yersinia enterocolitica*

8. Enrichment media is used to ___.

    a. inhibit pathogen growth
    b. favor normal flora growth
    c. inhibit normal flora growth
    d. all of the above

9. *Pseudomonas aeruginosa* is a particular threat to ___.

    a. rabbit hunters
    b. burn victims
    c. pregnant women
    d. professional painters

10. The true causative agent of "flu" (not influenza) is ____.

    a. *Yersinia pestis*
    b. a virus
    c. *Pasteurella multocida*
    d. *Haemophilus influenzae*

11. Both *Salmonella* and *Shigella* are _____.

    a. causative agents of typhoid fever
    b. true pathogens
    c. coliforms
    d. gram-positive bacilli

12. Enteric pathogens most often cause __ diseases.

    a. respiratory
    b. diarrheal
    c. skin
    d. sexually transmitted

13. New cases of pertussis have ___ in the United States since 1981.

    a. slowly declined
    b. rapidly declined
    c. rapidly increased
    d. slowly increased

14. Which of the following organisms is NOT a pathogen of concern as a potential bioterrorist threat?

    a. *Francisella tularensis*
    b. *Brucella abortus*
    c. *Bordetella pertussis*
    d. *Yersinia pestis*

15. K antigens are part of the bacterial ____ and are recognized by a host as foreign.

    a. capsule
    b. cell wall
    c. cytoplasm
    d. flagella

Applications of Chapter 20

1. A vaccine for *P. aeruginosa* was recently tested in cystic fibrosis patients. Why are these patients good candidates for this vaccine?

2. Why is serotyping pathogens such as *Salmonella* important from an epidemiology perspective?

Chapter 21 Miscellaneous Bacterial Agents of Disease

Building Your Knowledge

1. What type of cell wall do spirochetes have—gram-negative or gram-positive?

   Spirochetes are flagellated. Where would you find spirochete flagella?

   How do spirochetes move?

2. What is the causative agent of syphilis?

   Is this agent easy or difficult to grow in culture?

   How is syphilis transmitted and what is the infectious dose?

3. Differentiate between primary, secondary, and tertiary syphilis.

	Primary	**Secondary**	**Tertiary**
Time after exposure			
Major symptoms			
Treatment options?			

4. How is congenital syphilis contracted and what are the symptoms?

5. How is infection with *T. pallidum* diagnosed?

   How is syphilis commonly treated?

6. How are bejel, yaws, and pinta similar to syphilis? How are these diseases different?

7. What disease does *Leptospira interrogans* cause?

   How is this disease transmitted?

   How is leptospirosis diagnosed, treated?

   What preventative measures can be taken against the disease?

8. How is *Borrelia* transmitted to humans?

   What is the causative agent of relapsing fever?

   What are the symptoms of relapsing fever?

   How can relapsing fever be prevented or treated?

9. How is Lyme disease transmitted? What are endemic regions for Lyme disease?

   What are the early signs of infection with the Lyme disease spirochete?

   What organism causes Lyme disease?

   How can Lyme disease be prevented or treated?

   How is Lyme disease diagnosed?

10. Draw a vibrio.

11. How is cholera spread?

    What are the major symptoms of cholera and how is it treated?

12. What two vibros may be found in seafood?

    a. ______________________________

    b. ______________________________

13. What disease does *Campylobacter jejuni* cause?

    How is this disease transmitted?

    What are the symptoms and treatment for this infection?

    How is *Campylobacter jejuni* infection diagnosed?

    What other diseases do *Campylobacter* species cause?

14. What pathogen is the causative agent of stomach ulcers?

    How is this pathogen transmitted?

    How are ulcers now treated and why this course of action?

15. What diseases are caused by rickettsias?

16. How are rickettsial diseases transmitted?

17. What is the causative agent of typhus and how is it transmitted?

    What are the major symptoms of typhus?

18. What is Rocky Mountain spotted fever and what causes it?

19. What are the symptoms of Q fever and what is the causative agent of Q fever?

20. What is the causative agent of cat-scratch disease (CSD) and how can the disease be prevented?

21. Which diseases are caused by *Chlamydia* species?

    Why is azithromycin (an antibiotic that works intracellularly) a good choice to treat *Chlamydia* infections?

22. Which bacteria lack cell walls?

    Why is penicillin a poor choice to treat mycoplasmas and *Chlamydia*?

23. Draw a tooth, labeling the crown, root, enamel, and pulp cavity.

24. Which oral bacteria are commonly associated with dental caries?

25. Which is more serious, gingivitis or peridontitis, and why?

26. How are most dental diseases controlled?

Organizing Your Knowledge

Table 1. Disease Overview

Disease	**Causative Agent**	**Mode of Transmission**
Q Fever	a.	Airborne
Syphilis	*Treponema palladium*	g.
Yaws	b.	Invasion of skin
Leptospirosis	*Leptospira interrogans*	h.
Lyme disease	c.	Hard tick bite
Cholera	*Vibrio cholera*	i.
Ulcers	d.	Oral–oral, fecal–oral
Rocky Mountain spotted fever	*Rickettsia rickettsii*	j.
Cat-scratch disease	e.	Cat scratch
"Walking pneumonia"	*Mycoplasma pneumoniae*	k.
Epidemic typhus	f.	Lice
Ocular trachoma	*Chlamydia trachomatis*	l.

Self-Test: Vocabulary

1. _____ causative agent of syphilis
2. _____ mechanism by which spirochetes move
3. _____ causative agent of cholera
4. _____ microscopy technique used to diagnose spirochete infections
5. _____ early indication of syphilis is the appearance of a ____
6. _____ spirochete disease transmitted by infected animal urine
7. _____ painful, swollen, syphilitic tumors
8. _____ causative agent of Lyme disease
9. _____ rash at the site of a tick bite seen in most cases of Lyme disease
10. _____ mature oral biofilm found on teeth
11. _____ causative agent of stomach ulcers
12. _____ bacterial obligate intracellular pathogens that are transmitted by arthropod vectors
13. _____ *Coxiella burnetii* is the causative agent of ___
14. _____ cell wall-less bacteria that cause walking pneumonia
15. _____ bird-borne disease caused by *Chlamydiophila*

A. *Borrelia burgdorferi*
B. chancre
C. confocal
D. dark-field
E. endoflagellum
F. erythema migrans
G. gummas
H. *Helicobacter pylori*
I. leptospirosis
J. lice
K. mycoplasmas
L. ornithosis
M. plaque
N. Q fever
O. *Rickettsias*
P. *Treponema pallidum pallidum*
Q. *Vibrio cholera*

Self-Test: Multiple Choice

1. *Erythema migrans* is a characteristic of ____.
   a. relapsing fever
   b. cholera
   c. Lyme disease
   d. tertiary syphilis

2. Both *Campylobacter* and cholera are ____.
   a. vibrios
   b. arthropod-borne pathogens
   c. respiratory pathogens
   d. gram-positive

3. All spirochetes ___.
   a. cause disease
   b. are sexually transmitted
   c. are acid-fast
   d. are gram-negative

4. Mycoplasmas are __.
   a. all acid-fast
   b. always sexually transmitted
   c. lacking cell walls
   d. the causative agents of trench fever

5. Both gummas and a hard chancre are signs of ___.
   a. AIDS
   b. primary syphilis
   c. syphilis
   d. tertiary syphilis

6. Both *Bartonella* and *Coxiella* are ___.
   a. causes of Q fever
   b. bacterial pathogens
   c. viral pathogens
   d. water-borne pathogens

7. Chlamydial infection can lead to ___.
   a. blindness
   b. fevers and headache
   c. pelvic inflammatory disease
   d. all of the above

8. Treponemes, rickettsias, and chlamydias all ___.
   a. require cells for cultivation
   b. are transmitted sexually
   c. are vector-borne pathogens
   d. are energy parasites

9. The natural hosts and source for the bacteria that cause syphilis are ___.
   a. sheep and cows
   b. humans
   c. sheep only
   d. soil

10. The causative agent of walking pneumonia is ___.
    a. vibrio
    b. spirochete
    c. mycoplasma
    d. rickettsia

11. Elementary bodies and reticulate bodies are stages of ____.
    a. treponemes
    b. gummas
    c. rickettsias
    d. chlamydia

12. The causative agent of ulcers is ____.
    a. stress
    b. *Helicobacter pylori*
    c. *Vibrio vulnificus*
    d. *Campylobacter jejuni*

13. The life-threatening symptom of cholera infection is ___.
    a. secretory diarrhea
    b. lockjaw
    c. flaccid paralysis
    d. cholera is not a life-threatening disease

14. A disease transmitted by exposure to infected animal urine is ____.
    a. relapsing fever
    b. borreliosis
    c. leptospirosis
    d. yaws

15. Rocky Mountain spotted fever is caused by ____.
    a. a chlamydia
    b. a rickettsia
    c. exposure to contaminated water
    d. *Coxiella burnetti*

Applications of Chapter 21

1. Often when a person tests positive for syphilis, the physician will counsel the patient to get tested for chlamydia, gonorrhea, and HIV. Why?

2. Why is tertiary syphilis very rare in the United States today?

Chapter 22 The Fungi of Medical Importance

Building Your Knowledge

1. How are humans exposed to fungi?

2. Are most fungi pathogenic to humans? Explain.

3. How does thermal dimorphism contribute to fungal virulence?

4. How are most fungal pathogens transmitted?

   How do fungal epidemics commonly occur? Give a specific example.

5. How do most mycoses agents enter the human body?

6. List four separate fungal virulence traits.

   a. ______________________________

   b. ______________________________

   c. ______________________________

   d. ______________________________

7. Why is culturing fungus not the most common method to diagnose a fungal infection?

   How are fungal infections diagnosed? (List three rapid identification methods used.)

   Why is identification of a fungal infection (and distinguishing it from a bacterial infection) critical when treating immunocompromised patients?

8. How are fungal infections treated?

9. What three basic structures or processes are common targets for antifungal drugs?

   a. ______________________________

   b. ______________________________

   c. ______________________________

10. How is histoplasmosis contracted and who is at risk to contract it?

    What are the major symptoms of histoplasmosis?

    How does *Histoplasma capsulatum* evade the immune system?

    How is histoplasmosis diagnosed and treated?

11. What is valley fever?

    What activities are associated with outbreaks of the disease?

    What are the major symptoms of coccidioidomycosis?

    How is coccidioidomycosis diagnosed?

12. What is the causative agent of blastomycosis?

    Where is this disease endemic and how is it transmitted?

    What are the symptoms of blastomycosis and how is it diagnosed and treated?

13. What is rose gardener's disease and what is it caused by?

14. Are the causative agents of chromoblastomycosis and phaeohyphomycosis inherently virulent?

    Are these agents thermally dimorphic?

15. What area(s) of the body are affected by dermatophytoses?

16. List three dermatophyte genera and the diseases they cause.

17. Where may ringworm affect the human body?

    How is ringworm diagnosed and treated?

18. How do superficial mycoses differ from subcutaneous mycoses?

    Which are inflammatory infectious processes and which simply cause cosmetic problems?

19. What diseases are caused by *Candida albicans*?

    Who is at risk for developing a *Candida* infection and how is infection treated?

20. What body systems are commonly affected by cryptococcal infection?

How is *C. neoformans* treated?

Who is at risk for *C. neoformans* infection?

What is a potentially fatal result of systemic cryptococcosis?

21. How is cryptococcosis diagnosed and treated?

22. What is *Pneumocystis (carinii) jirovecci* infection?

How is *Pneumocystis carinii* infection spread?

How is it treated?

23. What diseases are associated with fungal allergens?

Organizing Your Knowledge
Table 1

Fungus	**Disease Caused**	**Identified by__**
Histoplasma capsulatum	d.	"Fish-eye" yeast cells
a.	Coccidioidomycosis	h.
Pneumocystis carnii	e.	Phagocytes with multiple cells—PCR test
b.	Ringworm	i.
Candidia albicans	f.	Pseudohyphae and light blue colonies on trypan blue agar
c.	Rose gardener's disease	j.
Cryptococcus neoformans	g.	Negative staining showing yeast cells with capsules

Self-Test: Vocabulary

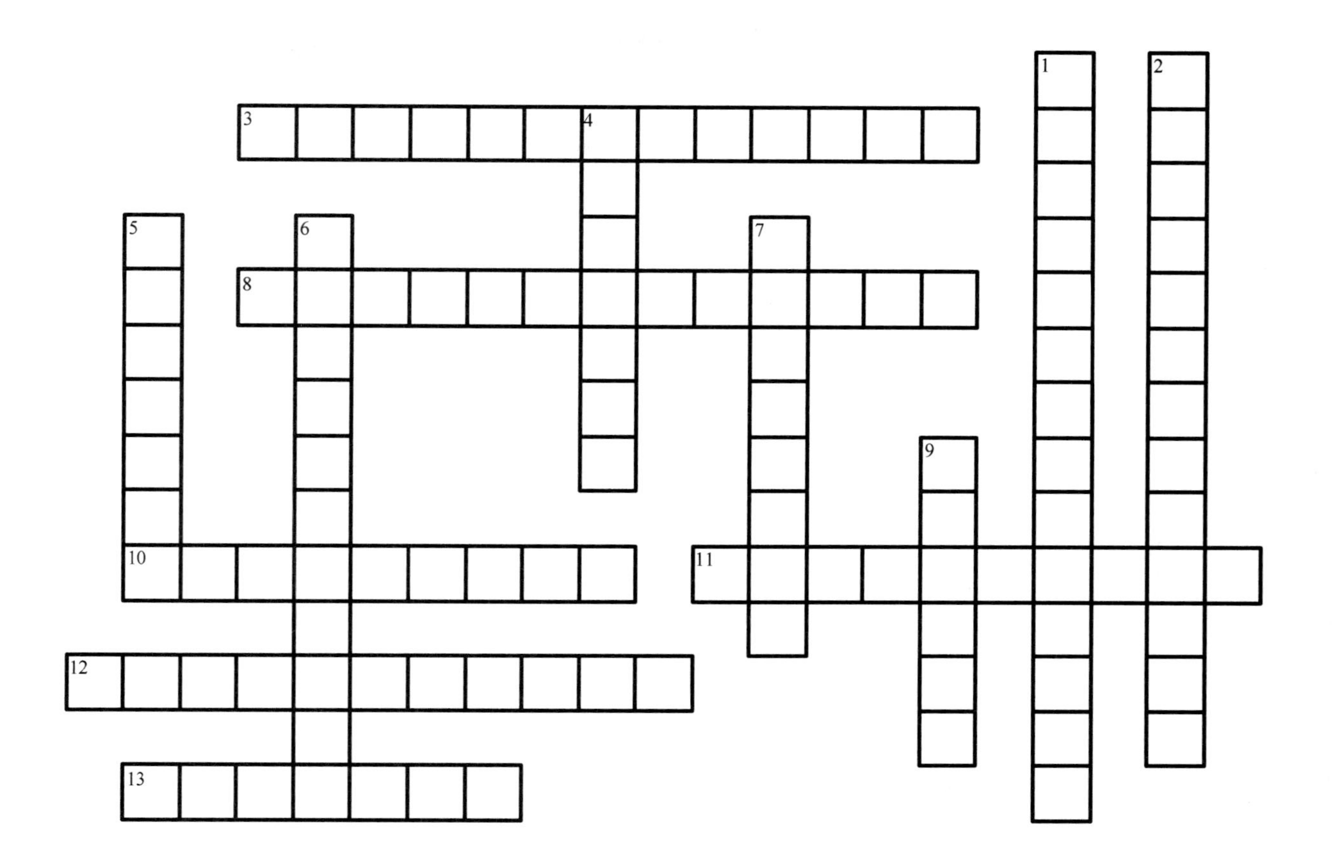

ACROSS

3 Chicago disease
8 fungus that cause ringworm and athlete's foot
10 a product of Aspergillus flavus that can contaminate stored grains
11 the ability to take two shapes
12 systemic anti-fungal that may be used to treat cryptococcosis
13 antifungal that disrupts plasma membranes

DOWN

1 rose-gardener's disease
2 pathogens that only infect the immunocompromised
4 fungal diseases
5 nodular growths seem with Valley Fever
6 common portal of entry for primary pathogens
7 topical antifungal used to treat Candida infections
9 oral candidiasis

Self-Test: Multiple Choice

1. Which of the following is NOT a common target of antifungal drugs?
   a. peptidoglycan
   b. plasma membranes
   c. cell division
   d. nucleic acid synthesis

2. The causative agent of California disease or San Joaquin valley fever is a(n) ______.
   a. *Histoplasma*
   b. *Coccidioides*
   c. form of ringworm
   d. *Blastomyces*

3. Most fungal species are ______.
   a. true pathogens and readily infect humans
   b. opportunistic pathogens and may infect humans
   c. nonpathogenic to humans
   d. procaryotic

4. *Cryptococcus neoformans* infection generally leads to __.
   a. loss of hair
   b. lost of fingernails
   c. meningitis
   d. pulmonary infection

5. Which of the following is a subcutaneous fungal infection?
   a. blastomycosis
   b. sporotrichosis
   c. dermatophytosis
   d. histoplasmosis

6. Dermatophytes and candida are both _____.
   a. nontransmissible fungi
   b. lethal respiratory pathogens
   c. communicable fungi
   d. types of ringworm

7. Dimorphic fungi are generally ____ that ___.
   a. decomposers: live in the soil
   b. pathogens: cause disease
   c. commensual organisms: form spores
   d. nonpathogens: are extinct

8. Pathogenic fungi are generally ___ at 30° C and ____ at human body temperatures.
   a. procaryotic: eucaryotic
   b. molds: yeasts
   c. yeasts: molds
   d. haploid: diploid

9. Histoplasmosis outbreaks are associated with exposure to ______.
   a. infected people
   b. bird droppings
   c. contaminated food
   d. rose bushes

10. When rapid identification of a fungus is required, which technique is NOT generally used?
    a. serology
    b. PCR
    c. hyphal morphology
    d. culturing fungal spores

Applications of Chapter 22

1. Commonly mild fungal infections, even after a diagnosis of an infection with primary pathogen, are left untreated. Why do you think this is the case?

2. There was a dramatic increase in the morbidity and mortality due to mycoses in the human population at the end of the 20th century. Why?

Chapter 23 The Parasites of Medical Importance

Building Your Knowledge

1. According to the World Health Organization (WHO), what percentage of all infections are caused by parasites?

2. What factors have increased the worldwide distribution of parasites in recent years?

3. What are the four groups of protozoan parasites?

    a. ______________________________

    b. ______________________________

    c. ______________________________

    d. ______________________________

4. How does a trophozoite differ from a cyst? Which is most often the infective body?

5. What is a karyosome and which stage of *Entamoeba histolytica*'s life cycle is it found in?

6. What are chromatoidals and which stage of *E. histolytica*'s life cycle are they found in?

7. What is the primary host for *E. histolytica*?

    Where is the incidence of *E. histolytica* infection the highest?

    What variables determine the severity of infection?

8. How does *E. histolytica* cause disease in humans and what may happen in the course of untreated amebiasis?

    How is amebic dysentery diagnosed?

9. Draw the *E. histolytica* life cycle. Label the cyst stage, encystment, and trophozoite stage. Indicate where trophozoites attach and multiply in the human body.

How can infection with *E. histolytica* be prevented?

10. What are two causative agents of amebic brain infection?

a. ______________________________________

b. ______________________________________

11. Where are these organisms normally found?

Why is early treatment of *Naegleria fowleri* infection crucial?

12. What type of protozoan is *Balantidium coli* and what disease does it cause in humans?

How is balantidiosis commonly spread ?

13. What are the four flagellated protozoans discussed in the text?

14. What disease is caused by *Trichomonas vaginalis* infection and how is it spread?

How common is trichomoniasis in the United States?

How is the disease diagnosed and treated?

15. What causes giardiasis?

How is *Giardia* transmitted?

How is giardiasis diagnosed?

How can giardiasis be prevented?

16. Why are trypanosomes and *Leishmania* species considered hemoflagellates?

How are hemoflagellates transmitted?

17. Draw the life cycle of *Trypanosoma cruzi*. Label the vector, human host, and the amastigote, promastigote, epimastigote, and trypomastigote stages.

Which stage is cystlike?

Which stage is the most complex?

Does most differentiation take place in the vector or human host?

18. What disease is caused by *Trypanosoma brucei*?

How does is this disease spread?

What are the major symptoms of *T. brucei* infection?

19. What disease is cause by *Trypanosoma cruzi* and how is this disease spread?

20. What are the major symptoms of Chaga's disease?

21. How is leishmaniasis transmitted?

    Which stage of the parasite's life cycle is transmitted to a human host?

    Where do Leishmania parasites convert to the amastigote stage?

    Differentiate between cutaneous leishmaniasis and systemic leishmaniasis.

22. Which three human pathogens are apicomplexans?

    a. ______________________________

    b. ______________________________

    c. ______________________________

23. Which four species collectively cause malaria?

    What percentage of the world's population lives in malaria-endemic regions?

    How many new cases of malaria are diagnosed each year?

    How many deaths are caused by malaria each year?

24. The life cycle of the malaria parasites has a sexual phase, carried out in the ______ and an asexual phase, carried out in the __________________.

25. Sporozoites enter liver cells, differentiate, and reproduce. How many merozoites are produced for every infected liver cell?

    Which stage of the parasite's life cycle is found in red blood cells?

    How do malarial parasites complete sexual reproduction?

    What are the symptoms and stages of malaria?

26. How does sickle-cell anemia protect against malaria?

    How is malaria treated?

    How may humans reduce the risk of becoming infected with malaria?

27. Why is developing an antimalarial vaccine so difficult?

28. What is *Toxoplasma gondii* and how widely distributed is the parasite?

    What is the primary reservoir for toxoplasmosis?

    How does the reservoir acquire the parasite?

    Which populations are particularly at risk to develop severe complications from toxoplasmosis?

    How can the disease be prevented?

29. What are the symptoms of *Cryptosporidium* infection?

    How is *Cryptosporidium* transmitted?

    How is infection with *Cryptosporidium* diagnosed and treated?

30. How is *Cyclospora cayetanensis* transmitted?

    What are the symptoms of infection?

31. What are the medically significant helminth worms?

32. Differentiate between intermediate, definitive, and transport hosts.

33. In most cases, how are humans exposed to helminth parasites?

34. Why is the presence of parasites in or on humans more accurately considered an infestation rather than an infection?

35. In what areas of the world do most helminth diseases occur?

How do certain cultural practices aid in the spread of helminth populations?

Are most helminth infections localized or systemic?

36. Why do intestinal worms often cause weight loss?

37. Which type of leukocyte is most responsible for eliminating worms?

38. How are helminth diseases diagnosed?

Why are most antihelminth drugs toxic to the host as well as the parasite?

39. How are intestinal nematodes different from tissue nematodes? Give examples of each.

40. How is ascariasis transmitted from one host to another?

What happens to the nematode after a human ingests *Ascaris* eggs?

41. What is the “hook” of a hookworm?

How are hookworms different from the other intestinal parasites discussed so far?

42. How do humans catch trichinosis?

Why are humans dead-end hosts to the infectious cycle?

43. Name two filarial worms and the diseases they cause.

44. What are the symptoms of elephantiasis and how are these symptoms related to the growth of the parasite in the human body?

45. What vector transmits river blindness?

46. Schistosomes have a complex life cycle with many stages. What stage infects snails?

Which stage infects humans?

What are the stages and symptoms of schistosomiasis?

What steps are being taken to control the spread of schistosomiasis?

47. How are lung and liver flukes transmitted?

48. Tapeworms have a simple structure of scolex and proglottids. Please draw a tapeworm, labeling the scolex and proglottids.

49. Describe taeniasis.

50. Why are arthropod control measures, such as DDT used to limit exposure to parasites and viral diseases?

Organizing Your Knowledge

Parasitic Agent	**Nature of Parasite**	**Disease Caused**	**Transmitted by__**
Balantidium coli	a.	Balantidiosis (dysentery)	n.
Cryptosporidium	Apicomplexan	h.	Contaminated water
Entamoeba histolytica	b.	Dysentery	o.
Giardia lamblia	Flagellate	i.	Contaminated water (zoonosis)
Leishmania tropica	c.	Cutaneous or systemic Leishmaniasis	p.
Naegleria fowleri	Amoeba	j.	Contaminated water
Onchocerca volvulus	d.	River blindness	q.
Plasmodium spp.	Apicomplexan	k.	Mosquito
Schistosoma spp.	e.	Schistosomiasis	r.
Trichinella spiralis	Nematode	l.	Undercooked pork
Trichomonas vaginalis	f.	Trichomoniasis	s.
Trypanosoma cruzi	Hemoflagellate	m.	Tsetse fly
Wurchereria bancrofti	g.	Elephantiasis	t.

Self-Test: Vocabulary

1. _____ disease common to hikers and campers drinking "clear mountain water"
2. _____ active feeding stage of a protozoan parasite
3. _____ amebic brain abscesses may be caused by ___
4. _____ ciliate parasite transmitted by pig feces
5. _____ sexually transmitted disease caused by an infectious flagellate
6. _____ causative agent of Chaga's disease
7. _____ flagellate that infects the blood
8. _____ trypanosome transmitted by the tsetse fly
9. _____ host where parasite matures to adult form
10. _____ causative agent of malaria
11. _____ parasite that is commonly found in cat feces
12. _____ disease that can be gotten by eating undercooked pork
13. _____ blood fluke with a snail intermediate host
14. _____ reproductive structures found in cestodes
15. _____ filarial worms may block lymph channels and cause ___

A. *Balantidium coli*
B. cyst
C. definitive
D. elephantiasis
E. giardiasis
F. hemoflagellate
G. intermediate
H. *Naegleria fowleri*
I. *Plasmodium* spp.
J. proglottids
K. *Schistosoma* spp.
L. *Toxoplasma gondii*
M. trichinosis
N. trichomoniasis
O. trophozoite
P. *Trypanosoma brucei*
Q. *Trypanosoma cruzi*

Self-Test: Multiple Choice

1. The active feeding state of a typical protozoan parasite is a ____.
   a. cyst
   b. hemoflagellate
   c. phlebotomine
   d. trophozoite

2. The least common protozoan infections are caused by ___.
   a. apicomplexans
   b. ciliates
   c. ameobas
   d. flagellates

3. Which of the following is NOT a pathogenic protozoan group?
   a. ciliates
   b. diatoms
   c. flagellates
   d. apicomplexans

4. Both roundworms and tapeworms are ___.
   a. ciliate protozoans
   b. helminths
   c. vector-borne parasites
   d. filarial worms

5. Arthropods are transmission vectors for all of the following EXCEPT __.
   a. malaria
   b. Chaga's disease
   c. Leishmaniasis
   d. trichinosis

6. The female *Anopheles* mosquito completes the ___ phase of the malarial parasite's life cycle.
   a. asexual
   b. sexual
   c. merozoite
   d. hemolytic

7. Why are most helminth diseases most accurately described as infestations, not infections?
   a. all helminths are hermaphroditic
   b. all helminths reproduce asexually
   c. adult helminths don't multiply and grow to maturity in a single host
   d. no helminths can inflitrate the blood

8. Water-borne parasites include ___.
   a. *Plasmodium* and *Trichomonas*
   b. *Giardia* and *Cryptosporidium*
   c. promastigotes and amastigotes
   d. *Legionella* and *Leishmania*

9. Both schistosomiasis and malaria are ___.
   a. caused by protozoan infections
   b. gastrointestinal illnesses
   c. blood diseases with liver involvement
   d. transmitted by mosquito bite

10. Both the causative agents for elephantiasis and schistosomiasis are ___.
    a. filarial worms
    b. blood flukes
    c. mosquito-borne
    d. parasites

11. Prevention of toxoplasmosis is best accomplished by ___.
    a. use of insect repellant
    b. avoiding snail-infested waters
    c. wearing shoes when walking
    d. proper hygiene around cats
    e. not eating raw pork

12. A protozoan flagellate disease that is sexually transmitted is ___.
    a. giardiasis
    b. leishmaniasis
    c. trichomoniasis
    d. trypanosomiasis

13. Ameboid disease may directly cause ____ in humans.
    a. dysentery and liver failure
    b. muscle fatigue and anemia
    c. sleeping sickness
    d. meningoencephalitis and dysentery

14. Which of the following is mismatched?
    a. *Ascaris*—roundworm
    b. amoeba—*Naegleria*
    c. *Trypanosoma*—helminth
    d. malaria—apicomplexan

15. The causative agent of malaria is ___.
    a. an apicomplexan
    b. a filarial worm
    c. a hemoflagellate
    d. a neurociliate

Applications of Chapter 23

1. How may antibiotics, such as tetracycline, be helpful in treating a parasitic disease, such as river blindness?

2. If you were designing a strategy to vaccinate against malaria, would you target sporozoites or merozoites? Explain.

Chapter 24 Introduction to Viruses That Infect Humans: The DNA viruses

Building Your Knowledge

1. How are viruses significantly different from bacteria, fungi, and protozoans?

2. How are animal viruses divided into families?

   Are most DNA viruses generally single-stranded or double-stranded? Are most RNA viruses single- or double-stranded?

   Where does the envelope of an animal cell virus come from?

3. Why are certain viruses associated with certain cell types (e.g., hepatitis and liver cells)?

4. How do humans develop immunity to viruses?

5. How do chronic infections differ from latent infections?

6. What are oncogenic viruses?

7. Name two viruses with teratogenic effects.

8. List the six groups of DNA viruses that infect humans, indicating whether the virus is enveloped or naked and if it is single-stranded or double-stranded.

9. Describe the poxviruses. Where do these poxviruses proliferate in the human body? What causes pox to form?

10. What is the difference between variola and vaccinia?

11. How serious is smallpox infection? How was it eradicated and when?

    How is smallpox transmitted?

    What are the symptoms of smallpox?

12. What is molluscum contagiosum and how is it contracted?

    How is it treated?

13. Which mammalian poxviruses can cause disease in humans?

14. List the pathogenic herpesviruses and the diseases they cause.

Herpes Virus	Disease Caused
a.	Fever blisters, genital infections
Varicella-zoster viruse (VCV)	d.
b.	Salivary glands and viscera
Epstein-Barr (EBV)	e.
c.	Roseola
Herpesvirus-8	f.

15. Compare and contrast herpes simplex type 1 and type 2 infections.

How are the two viruses spread?

What are the symptoms and complications of infection?

How do each become latent and how do recurrent attacks occur?

How does herpes simplex 1 manifest itself in children?

How is herpes simplex 2 transmitted and what are the major symptoms of the disease?

How are herpes infections diagnosed and treated?

What is whitlow and who is in danger of contracting it?

16. How are the causative agents for chickenpox and shingles related to one another?

Can someone catch chickenpox from someone with shingles?

What pattern do the lesions associated with shingles follow?

How are shingles activated?

How can shingles and chicken pox be prevented?

17. What is CMV?

Who is at risk of serious complications with CMV infection and what are the symptoms of congenital CMV?

Who is at risk for disseminated CMV?

What problems have been encountered to develop a vaccine for CMV?

18. How are Burkitt's lymphoma and mononucleosis related?

19. How does EBV infection manifest itself in developing countries?

How does EBV infection manifest itself in industrialized countries?

20. List three separate illnesses associated with EBV infection.

    a. ______________________________

    b. ______________________________

    c. ______________________________

21. What diseases are caused by HHV-6?

22. If cancer patients are seropositive for herpesviruses, does that indicate the virus caused the cancer? Why or why not?

23. What organ do hepadnaviruses infect? What disease(s) do they cause?

24. How many different viruses are known to cause hepatitis?

    Are these viruses related to one another?

25. How is hepatitis B spread?

    What is the range of HBV symptoms?

    How can hepatitis B virus infection be prevented?

26. What diseases are associated with adenovirus infection?

    How are adenoviruses transmitted from person to person?

27. What diseases are caused by papillomaviruses?

Why has there been recent interest in a HPV vaccine?

28. What one group of pathogenic DNA viruses are single-stranded?

What diseases do these viruses cause in humans?

What diseases do these viruses cause in animals?

Organizing Your Knowledge

Virus	Describe Virus	Disease Caused	Prevention
Adenovirus	Naked, double-stranded DNA	f.	Polyvalent vaccine used in military
Cytomegalovirus	a.	Opportunistic infection in neonates	k.
Epstein-Barr	Enveloped double-stranded DNA	g.	Avoid contact
HBV	b.	Hepatitis B virus	l.
HSV I	Enveloped double-stranded DNA	h.	Avoid contact—treat with acyclovir or valacyclovir
HSV II	c.	Herpes simplex II—genital herpes	m.
Papillomavirus	Naked double-stranded DNA	i	Vaccine for HPV
Parvovirus	d.	Fifth disease (humans), animal diseases	n.
Varicella-zoster	Enveloped double-stranded DNA	j.	Attenuated vaccine
Variola	e.	Smallpox	o.

Self-Test: Vocabulary

1.	_____ any process or agent that causes developmental disturbances and fetal damage	A. cervical cancer
2.	_____ causative agent of shingles also causes _____	B. chickenpox
3.	_____ causative agent of smallpox	C. double-stranded
4.	_____ both CMV and Epstein-Barr cause ___	D. host membrane
5.	_____ herpes viruses all exhibit this form of persistent infection	E. jaundice
6.	_____ HHV-6 causes ____ in infants	F. latency
7.	_____ human papilloma virus is one of the causative agents of ___	G. molluscum contagiosum
8.	_____ local infection of herpes, deep-set, painful and itchy abscess	H. monkeypox
9.	_____ most DNA viruses that infect humans are ___	I. mononucleosis
10.	_____ pox disease with smooth, waxy skin lesions	J. oncogenic
11.	_____ source of a viral envelope	K. parvoviruses
12.	_____ the only single-stranded DNA viruses that infect humans	L. roseola
13.	_____ virus used to prevent smallpox	M. single-stranded
14.	_____ viruses that can cause cancer are ____	N. teratogenic
15.	_____ yellow tinge to skin and eyes—generally a sign of liver damage	O. vaccinia
		P. variola
		Q. whitlow

Self-Test: Multiple Choice

1. Which of the following is NOT a sexually transmitted disease?
   a. papilloma
   b. variola
   c. hepatitis C
   d. all of the above are transmitted primarily through sexual contact

2. Adenoviruses are ___
   a. single-stranded RNA viruses
   b. double-stranded DNA viruses
   c. single-stranded DNA viruses
   d.single-stranded RNA viruses

3. Both vaccinia and variola are ____.
   a. poxviruses
   b. parvoviruses
   c. herpesviruses
   d. adenoviruses

4. Most DNA viruses are assembled in the ____ of an animal cell.
   a. cytoplasm
   b. Golgi apparatus
   c. mitochondria
   d. nucleus

5. Parvoviruses are unusual because they ____.
   a. cause coldlike symptoms
   b. are double-stranded DNA viruses
   c. have both DNA and RNA in the same viral particle
   d. are single-stranded DNA viruses
   e. infect only humans

6. Tetragenic viruses cause ___ and include ___.
   a. cancer: smallpox
   b. heart attacks: HIV
   c. birth defects: rubella
   d. meningitis: meningococcus

7. The Epstein-Barr virus causes ____.
   a. shingles
   b. birth defects
   c. hepatitis
   d. mononucleosis

8. Viral hepatitis occurs ___.
   a. as a result of many different viral infections
   b. as the result of a single viral infection
   c. only in humans
   d. only in infants

9. The smallpox virus reproduces in ___ cells primarily.
   a. liver
   b. epidermal
   c. nerve
   d. muscle

10. Most herpes viruses cause ___.
    a. cold sores to form
    b. latent viral infections that can recur
    c. AIDS
    d. cold and flulike symptoms

11. Chickenpox is caused by a(n) ____.
    a. adenovirus
    b. poxvirus
    c. herpesvirus
    d. cytomegalovirus

12. Most DNA viruses that are human pathogens are ___.
    a. single-stranded
    b. double-stranded
    c. there are no DNA viruses that are human pathogens
    d. triplex

13. Both mononucleosis and Burkitt's lymphoma are caused by ___.
    a. RNA viruses
    b. naked DNA viruses
    c. single-stranded DNA viruses
    d. Epstein-Barr virus

14. If a child has never had (or been immunized against) chickenpox, he or she can "catch" chickenpox from someone with shingles.
    a. True
    b. False

15. Which populations are at risk for developing serious complications with herpesvirus infections?
    a. neonates
    b. transplant patients
    c. AIDS patients
    d. all of the above

Applications of Chapter 24

1. Why are vaccines often the most effective defense against viral infection?

2. Viruses have only recently been associated with cancer, to the extent that there is now a vaccine against human papilloma virus that prevents cervical cancer. Why is it difficult to link viruses and cancer development?

Chapter 25 The RNA Viruses of Medical Importance

Building Your Knowledge

1. How many groups of RNA viruses cause disease in humans?

   How are these viruses separated?

2. Draw an orthomyxovirus. Label the hemagglutinin and neuraminidase, envelope, and RNA molecules.

3. How do hemagglutinin and neuraminidase contribute to the virulence of influenza virus?

4. Differentiate between antigenic shift and antigenic drift.

   Would influenza be as capable of antigenic shift if it had a nonsegmented genome? Why or why not?

   Why is it necessary to get a flu shot every year?

5. How are flu viruses named?

   Which flu pandemic of the 20th century had the greatest worldwide death toll?

6. What antiviral drugs are available to treat influenza?

7. How are the influenza vaccines made every year?

8. How does Flu-mist differ from the flu shot?

9. What disease is caused by hantaviruses and how is it spread?

10. What are the three major paramyxoviruses and what diseases do they cause?

Paramyxoviruses	**Disease Caused**
a.	Mumps and parainfluenza
Morbillivirus	c.
b.	Respiratory syncytial virus

How are all three paramyxoviruses spread?

11. What are the symptoms of parainfluenza virus infection in children and how are they treated?

Who is most susceptible to infection with parainfluenza virus?

12. How is the mumps virus transmitted?

Where does the virus multiply early in an infection?

How is mumps diagnosed and treated?

13. What is the natural reservoir for measles?

14. How is measles infection spread?

When are people who contract measles contagious?

How is measles diagnosed and treated?

How is measles infection prevented?

15. What are the symptoms of RSV in adults and older children?

What are the symptoms of RSV in infants?

Who is at the most risk to develop serious complications from RSV infection?

16. Which rhabdovirus is most concerning to human medicine?

What is the prognosis for untreated rabies?

How do humans contract rabies?

Where does rabies multiply early in an infection?

Where is the virus found in late stages of an infection?

Compare and contrast the furious and dumb forms of rabies.

Once rabies has been diagnosed, what is the typical course of treatment?

How is this unusual?

17. Describe a coronavirus.

What emerging disease is caused by a coronaviruse?

18. Describe a togavirus.

19. What is rubella and how is it spread?

Describe the two clinical forms of rubella.

How is rubella diagnosed?

How is rubella prevented?

20. How are arboviruses transmitted?

21. What factors influence the distribution and frequency of arboviral infections?

22. What are the arboviruses that cause encephalitis?

    Are humans the primary host for these viral pathogens?

23. What do yellow fever and Dengue fever have in common?

    How is yellow fever spread?

24. How do retroviruses differ from most RNA viruses?

25. Draw an HIV viral particle, labeling the envelope, spikes, and RNA.

26. What does the HIV blood screening test measure?

27. How can HIV be contracted?

28. How does HIV multiply within immune cells?

29. What are the AIDS-defining illnesses?

30. Why have congenital HIV infections dramatically decreased in recent years?

31. How is HIV transmitted?

32. How and why do T cells die during an HIV infection?

How do HIV viral particles enter the brain?

33. How can a person get a false-negative HIV test?

34. Please describe three separate anti-AIDS drugs and the viral life cycle segment they target.

Reverse transcriptase inhibitors:____

Protease inhibitors:____

Fusion inhibitors:____

35. Name two different (non-HIV) retroviral agents and the diseases they cause.

36. What type of virus causes polio?

37. How is polio transmitted?

    How does infection with a polio virus progress?

38. Differentiate between the Salk and Sabin vaccines. Which is live? Which is an inactivated vaccine? What are the advantages and disadvantages of each?

39. What is hepatitis A virus and how is it related to hepatitis B virus?

40. What group of viruses is the cause of the common cold?

    Why are rhinoviruses restricted to growth in the nose and upper respiratory areas?

    How may people minimize the spread of colds?

41. How are rotavirus and Norwalk virus similar?

    Who typically gets rotavirus? How is severe rotavirus infection treated?

42. How is a prion different from a viral particle?

What class of diseases is known to be caused by prions?

How does prion infection cause disease?

Organizing Your Knowledge

Virus	**Viral Group**	**Transmitted by ___.**
Common cold	Picornavirus	g.
Hepatitis A	a.	Fecal/oral routes
HIV	Retrovirus	h.
Influenza virus	b.	Droplet/aerosol
Measles	Morbillivirus	i.
Mumps	c.	Droplet—mucus and saliva
Polio	Picornavirus	j.
Rabies	d.	Animal bite
Rotavirus	Reovirus	k.
RSV	e.	Droplet/fomite
SARS	Coronavirus	l.
Yellow fever	f.	Mosquito

Self-Test: Vocabulary

ACROSS

5 labored breathing with a hoarse cough seen in young children with parainfluenza virus

7 zoonotic rhabdovirus that is almost always fatal if left untreated

9 infectious protein particle

11 German measles

12 minor change in antigen profile due to mutations

13 viral infection of parotid glands

14 causative agent of the common cold

15 vaccine for polio that is an attenuated virus given orally

16 double-stranded RNA virus that causes gastroenteritis

DOWN

1 influenza protein that binds to host cell receptors of the respiratory mucosa

2 having an affinity for the nervous system

3 viral disease characterized by Koplik's spots

4 antigenic changes due to recombining of different viral strains

6 hemorrhagic fever transmitted by rodents that causes pulmonary collapse

8 group of arthropod-borne viruses

10 virus that converts RNA to DNA

12 difficulty in breathing

Self-Test: Multiple Choice

1. Arboviruses are all___.
   a. zoonotic diseases
   b. arthropod-borne
   c. RNA viruses
   d. all of the above
   e. none of the above

2. In nature, Yellow fever and Dengue Fever are primarily transmitted by ___.
   a. mosquitos
   b. air droplets
   c. blood products
   d. unsafe sexual practices

3. How is poliovirus spread throughout a population?
   a. air-borne routes
   b. BSE-contaminated meat
   c. fecal–oral contamination
   d. blood and other bodily fluids

4. HIV is the only retrovirus that infects humans.
   a. True
   b. False

5. Mumps, polio, and rubella are all ___.
   a. RNA viruses
   b. zoonotic diseases
   c. spread through fecal-oral transmission
   d. enveloped viruses

6. The difference between HIV infection and AIDS is _______.
   a. there is a different causative agent
   b. AIDS patients are positive for anti-HIV antibodies, HIV patients are not
   c. AIDS patients have circulating virus, HIV patients do not
   d. AIDS is associated with decreased T cell counts and indicator conditions

7. Most RNA viruses are ___ and assemble in the __.
   a. double-stranded: nucleus
   b. single-stranded: nucleus
   c. double-stranded: cytoplasm
   d. single-stranded: cytoplasm

8. Kaposi sarcoma is caused by HIV infecting muscle cells.
   a. True
   b. False

9. If influenza had a nonsegmented genome, it would __.
   a. be related to the prion diseases
   b. not be infectious
   c. be less likely to show antigenic shift
   d. be a DNA virus

10. Hepatitis A is caused by ___.
    a. a retrovirus.
    b. a poliovirus
    c. a picornavirus
    d. a togavirus

11. HIV antiviral drugs target __.
    a. proteases
    b. reverse transcriptase
    c. the fusion of viral particles with cells
    d. all of the above

12. All retroviruses produce _.
    a. hemagglutinin
    b. parainfluenza
    c. reverse transcriptase
    d. spongiform particles

13. The causative agent of severe acute respiratory syndrome (SARS) is _.
    a. an influenza virus
    b. a pneumovirus
    c. a rhabdovirus
    d. a coronavirus

14. Rabies is an unusual infection because it ____.
    a. is untreatable
    b. responds to active immunization after exposure
    c. can't be spread between species
    d. is a viral infection that responds to antibiotics

15. HIV can be transmitted by mosquito bite.
    a. True
    b. False

Applications of Chapter 25

1. Given the typical process of mass-producing flu vaccine, why is the prospect of vaccinating against avian influenza—H5N1 a difficult one?

2. What is the likelihood of another 1918-like pandemic of influenza occurring in the 21st century?

3. Why is developing a vaccine against the common cold highly unlikely?

Chapter 26 Environmental and Applied Microbiology

Building Your Knowledge

1. Differentiate between applied and environmental microbiology.

2. Draw the levels of the biosphere as they relate to one another. Label the biosphere, hydrosphere, lithosphere, and atmosphere. There will be overlap.

3. How is a community different from a population?

4. Which organism would likely have a broader niche, a scavenger or a nitrogen fixer?

5. Are producers autotrophic or heterotrophic?

   What do producers produce and how do they produce it?

6. What would happen to an ecosystem that lost its decomposers?

7. Does the energy available to consumers increase or decrease as you move from primary to secondary and from secondary to tertiary consumers?

8. What five traits do most biogeochemical cycles share?

   a. ____________________

   b. ____________________

   c. ____________________

   d. ____________________

   e. ____________________

9. What do geomicrobiologists study?

10. Which atmospheric cycle is most closely associated with living systems, energy transformation, and trophic patterns?

11. What process fixes carbon dioxide into an organic form of carbon (sugar)?

    How is carbon dioxide released from living processes?

    What gas do methanogens release?

12. Why are photosynthetic pigments crucial to photosynthesis?

    What happens during the light-independent reactions of photosynthesis?

13. How is ATP synthesized during photosynthesis and when is it synthesized?

14. How does anoxygenic photosynthesis differ from oxygenic photosynthesis?

15. What organisms are able to fix nitrogen? Are plants able to fix nitrogen?

16. What are the nitrifying bacteria?

    What are the detrifying bacteria?

17. How is sulfur used in living organisms?

18. How can *Thiobacillus* survive in environments with few complex organic nutrients?

    What role does *Thiobacillus* play in the phosphorus cycle?

19. What role does phosphorus have in living organisms?

Why is phosphorus a major component of fertilizers?

What effect does excess phosphorus have on the hydrosphere?

20. What is bioamplification?

21. What is humus?

Both bogs and tropical forests have nutrient-poor soil, but for different reasons. Why are tropical forest soils generally poor?

22. What is the rhizosphere?

23. How are mycorrhizae beneficial to plants?

24. Draw a hydrologic cycle, labeling transpiration, precipitation, aquifers, surface water, and clouds.

25. What two problems are acquifers currently facing?

26. Draw a lake in cross-section. Label the photic, profundal, and benthic zones. Label the littoral and limnetic zones as well.

27. What is thermal stratification?

28. What role do bacteriophage play in aquatic environments?

29. What are the most prominent human pathogens that are water-borne?

30. Why are coliforms good indicator organisms, but coliform bacteriophages are not?

31. Compare and contrast standard plate counts with membrane filter methods.

32. What is an acceptable level of fecal coliforms in drinking water?

33. How is drinking water purified? Diagram the process from water source to kitchen sink.

34. How is sewage purified and what role do microbes play in the process? Diagram the process from kitchen sink to waterways.

35. How do industrial microbiologists define fermentation?

36. Give examples of beneficial actions of microbes on food and detrimental actions of microbes on food.

37. Do the same microbes that spoil food cause food-borne infections?

38. How is making bread similar to the process of making beer?

39. Diagram the process of beer making from ingredients to finished product.

40. Why is most beer between 3% and 5% alcohol?

41. Fill in the following chart.

Item Fermented	**Product**
Potatoes	c.
a.	Whiskey
Corn mash	d.
b.	Brandy

42. How is cheese made? Where does rennin come from and what is it used for?

43. How does a food-borne infection differ from a food-borne intoxication?

44. How many food-borne illnesses occur each year in the United States?

45. Why can utensils be treated with UV radiation, but food is treated with gamma irradiation?

Does irradiated food become radioactive?

46. How do primary and secondary metabolites differ from one another?

47. How may industrial microbiologists increase the amounts of a desired end product?

48. Draw the process of growing organic substances in a fermentor and the mass production of those substances.

49. How does a continuous system differ from batch fermentation?

50. What was the first mass-produced antibiotic?

Organizing Your Knowledge

Please indicate with an X in the appropriate column(s), which branch of microbiology each area of study belongs to.

Area of Study	Applied Microbiology	Environmental Microbiology
a. Bioremediation		
b. Water microbiology		
c. Food microbiology		
d. Energy pyramids		
e. Trophic structures		
f. Food webs		
g. Biogeochemical cycles		
h. Biomagnification		
i. Sewage treatment		
j. Production of enzymes		
k. Mass Production of drugs		
l. Production of vaccines		
m. Chemical cycling		

Self-Test: Vocabulary

1. _____ bacteria that live in the roots of legumes and fix nitrogen
2. _____ science that studies interactions between microbes and their environment
3. _____ climatic regions with dominant plant forms, temperatures, and precipitation
4. _____ deep groundwater source in which water circulates slowly
5. _____ drinkable water
6. _____ living or dead organisms that occupy an organism's habitat are ___ factors
7. _____ microbe that generates methane as part of its metabolism
8. _____ nutrient-poor ecosystems
9. _____ physical location to which an organism has adapted
10. _____ process by which pollutants accumulate in living tissue
11. _____ sewage treatment stage where microbes digest organic waste
12. _____ slowly decaying organic litter found in soil
13. _____ the water portion of global ecosystems
14. _____ using light energy to split water
15. _____ using microbes to break down toxins in water or soil

A. aquifer
B. bioamplification
C. biome
D. bioremediation
E. biotic
F. habitat
G. humus
H. hydrosphere
I. methanogen
J. microbial ecology
K. oligotrophic
L. photolysis
M. potable
N. primary phase
O. *Rhizobium*
P. secondary phase

Self-Test: Multiple Choice

1. The same organisms that spoil food will often cause disease in humans.
   a. True
   b. False

2. Rennin is an important part of ___.
   a. industrial fermentors used to make penicillin
   b. bread making
   c. beer making
   d. cheese making

3. Whey is a ________.
   a. normal product of yogurt culture
   b. by-product of wine making
   c. toxic metabolite from cheese making
   d. normal product of cheese production

4. In photosynthesis, the ATP that is produced ____.
   a. is used to fix carbon dioxide to sugar
   b. is used by the cell
   c. powers photosystem II
   d. powers photosystem I

5. Which of the following is NOT a commonly used method to preserve food?
   a. radiation
   b. vinegar
   c. salt
   d. milk proteins

6. Plankton often live in the ____ zone of a lake.
   a. benthic
   b. profundal
   c. abyssal
   d. photic

7. Different soil levels differ greatly on their _____.
   a. oxygen availablility
   b. microbial flora
   c. acidity
   d. all of the above

8. Microbes are a crucial part of the ____ stage of sewage treatment.
   a. primary
   b. secondary
   c. tertiary
   d. chlorination

9. Primary metabolites are___.
   a. generally what's harvested from a fermentor
   b. essential to a microbe's function
   c. generally toxic to humans
   d. are synthesized during log phase of a growth cycle

10. Bioamplification contributes to the ___.
    a. accumulation of decomposers in the soil
    b. accumulation of toxins in the food chain
    c. greenhouse effect
    d. depletion of ozone

11. The soil in bogs is generally nutrient-poor because ___.
    a. microbes decompose nutrients too quickly
    b. not enough plant life to form nutrients
    c. too much oxygen for anaerobes to work properly
    d. high acid content and slow microbial decomposition

12. Filtration of water for drinking is an important way to ____.
    a. remove bacteriophage
    b. remove protozoan cysts
    c. chemically disinfect water
    d. irradiate fresh water

13. If a person catches salmonella from eating undercooked chicken, that person has a food-borne _____.
    a. infection
    b. intoxication
    c. spoilage
    d. fermentation

14. The form of pasteurization that will result in the longest shelf life is ___.
    a. high temperature, short time
    b. low temperature, long time
    c. ultra high temperature
    d. ultra low temperature

15. Excessive phosphorus in runoff will often lead to ___.
    a. oligotrophication of lakes
    b. a spike in giardiasis cases
    c. eutrophication of lakes
    d. no deleterious effects—the environment absorbs the excess

Applications of Chapter 26

1. There has been a great deal of interest in bioremediation in recent years. What are the advantages of bioremediation over chemical methods?

2. There are microbes that are able to chemically convert gold and other heavy metals into soluble form. If you were an industrial microbiologist, what traits would you look for in a "biomining" microbe?

Foundations of Microbiology Student Study Guide
Answer Key

Chapter 1

Building Your Knowledge

1. See table 1.1, bacteriology, mycology, virology, etc.
2. Microbes are small and have a short generation time; however, they cannot be directly seen.
3. See figure 1.1. a. origin of Earth; b. procaryotes appeared; c. eucaryotes appeared; d. reptiles; e. cockroaches/termites; f. mammals; g. humans.
4. Microbes have a ubiquitous distribution. Every location that has been sampled has microbes. Procaryotes are more numerous than eucaryotes.
5. Photosynthesis is the fixing of carbon from an inorganic ($C0^2$) to an organic (sugar) form. Decomposition returns molecules to a simpler form from a complex organic form.
6. Genetic engineers manipulate the genes of both procaryotes and eucaryotes to produce a gene product of interest or genetically modified organism. Drugs such as insulin, hormones such as human growth factor, and GMO such as Bt corn are the product of genetic engineering.
7. Bioremediation uses microbes to clean up toxic pollutants in the environment.
8. Any agent that can cause disease is considered a pathogen. A pathogen may be bacterial, viral, protist, fungal or prion in nature.
9. Both procaryotes and eukaryotes are cellular forms. Viruses are not cells.
10. Procaryotes lack a nucleus and membrane-bound organelles and they are on average considerably smaller than eucaryotes.
11. Antonie van Leeuwenhoek was an amateur microscopist who was the first person to see procaryotic cells. Louis Pastuer disproved spontaneous generation, developed a vaccine for rabies and the process of pasteurization. Robert Koch developed a method to determine which organism is a pathogen and cause of a particular disease. Joseph Lister started the process of aseptic surgery.
12. The three domains of life are Archaea, Bacteria and Eucarya. Bacteria and Archaea are procaryotic.
13. *Mycobacterium leprae.*

Self-Test:Vocabulary

1. L	2. M	3. I	4. F	5. E	6. D	7. H
8. J	9. G	10. B	11. K	12. C	13. A	

Self-Test: Multiple Choice

1. e	2. b	3. e	4. a	5. a
6. c	7. b	8. d	9. e	10. d
11. d	12. c	13. b	14. c	15. b

Applications of Chapter 1

Answers will vary. Procaryotes can live in extreme environments, well outside what is survivable for eucaryotes in terms of pH, temperature, salt, radiation, or oxygen levels. The relatively temperate conditions required for eucaryotes to thrive would not likely be found in extraterrestrial environments.

Chapter 2

Building Your Knowledge

1. An atom.
2. See figure 2.1. Helium has both a proton and a neutron in the nucleus and two electrons.
3. Atomic number is the number of protons in an element. Atomic mass is the number of protons and neutrons in an element. Adding electrons will change an atom's charge, not its mass or its atomic number. An uncharged atom with 12 protons would have 12 electrons.
4. Isotopes.
5. Electrons.
6. The atom can accept two electrons and participate in two bonds.
7. Electrons are shared in covalent bonds and are given (or taken) in ionic bonds.
8. Water is a polar molecule. Nonpolar molecules have their electrons shared equally, so no partial charge exists on either end of the molecule.
9. Both cations and anions are charged molecules; however cations are positively charged and anions are negatively charged.
10. Electrons are not shared or given in hydrogen bonds. Polar molecules will be attracted to one another since they have positive and negative partial charges.
11. Molecular formulas will give the elements present in a compound, but not how they are arranged. Structural formulas (see figure 2.11) indicate how the atoms are arranged. Glucose and fructose have the same molecular formula, but different structural formulas.
12. $A + B \longrightarrow C$ (synthesis reaction).
13. Water is the solvent, the powder is the solute and the lemondae is an aqueous solution.
14. Amphipathic molecules have a polar and nonpolar end. Polar molecules dissolve in water and are hydrophilic. Nonpolar molecules do not and are hydrophobic.
15. 5%. If you dissolve 5 grams in 50 ml of water, you would have a 10% solution, which is more concentrated.
16. Acids release H+, bases release OH-. A solution of a pH of 3 has more H+ than a solution of pH of 5 and is more acidic.
17. Carbon has four electrons in its valence shell so can participate in four different covalent bonds. This makes it ideal for making large macromolecules, such as proteins, nucleic acids, or lipids.
18. Functional groups are molecular groups that make a larger molecule able to react with enzymes or other molecules. Examples include the amino end of an amino acid, the phosphate end of a nucleotide or the hydroxyls found in sugar molecules.
19. Nucleic acids, proteins, carbohydrates, lipids.
20. The -ose ending would indicate the molecule is a sugar (carbohydrate).
21. Dehydration synthesis is the removal of water to join monomers. Carbohydrates are built by dehydration synthesis and degraded by hydrolysis.
22. Oil and water don't mix because oils are nonpolar and water is a polar molecule. Lipids are a major part of the plasma membrane and if they dissolved in water they could not serve as a barrier in aqueous solutions.

23. Fatty acids and glycerol are lipid subunits. Lipases are enzymes that break apart lipids into their subunits.
24. See figure 2.20. Hydrophobic tails are inside the bilayer, hydrophilic heads are outside.
25. Membranes must form a barrier between the inside and outside of the cell. They must let nutrients into the cell and wastes out of the cell.
26. See figure 2.21. Amino group is the NH^2, carboxyl is the COOH. A dehydration synthesis reaction removes water to join the carboxyl end to the amino end of two different amino acids.
27. Covalent.
28. Enzymes, antibodies, receptors, toxins, and structural proteins.
29. DNA and RNA are polymers of phosphate-linked 5-carbon sugars that have nitrogenous bases attached to their first carbon of the five carbon sugar. DNA is generally double-stranded and has deoxyribose. RNA is generally single-stranded and has ribose as its 5-carbon sugar.
30. See figure 2.23.
31. Purines are adenine and guanine and have two rings containing nitrogen and carbon. Pyrimidines are thymine cytosine, and uracil and have one ring.
32. Phosphodiester. Hydrogen.
33. Transfer RNA (carrier of amino acids), messenger RNA (copy of the gene), ribosomal RNA (component of ribosomes).
34. ATP is a nucleotide, like the nucleotides that make up DNA and RNA. However, it is used to store chemical energy for a cell.

Organizing Your Knowledge

Table 1

a. Sugar molecules
b. Starch or glycogen
c. Lipids
d. Membrane integrity
e. Phospholipids in plasma membrane
f. Amino acids
g. Structural, defense, catalyzing reactions
h. Nucleic acids
i. Transmission of information

Table 2

a. Semisolid media
b. Chitin
c. Polymer of glucosamine
d. Bacterial cell walls
e. Glycans linked to peptide fragments
f. Cellulose
g. Structural support
h. Plants or animals
i. Storage of energy (carbon)

Table 3

a. Association of polypeptides with each other
b. Primary
c. Hydrogen bonding between the amino acids in a chain (e.g. alpha helix or beta sheet)
d. Tertiary

Table 4

1. Membrane components
2. Cholesterol
3. Inflammation and allergy

Self-Test: Vocabulary: ACROSS: 2 lipid, 5 proton, 6 covalent, 7 acid, 13 carbohydrates, 14 glycosidic, 15 ribose; DOWN: 1 secondary, 3 polar, 4 organic, 8 enzyme, 9 peptide, 10 valence, 11 solvent, 12 pentose

Self-Test: Multiple Choice

1. a	2. e	3. e
4. d	5. d	6. c
7. b	8. d	9. c
10. d	11. b	12. b
13. d	14. e	15. c

Applications of Chapter 2

Carbon has four valence electrons, so is able to participate in four different bonds. This makes it well suited to form large macromolecules, such as sugars, lipids and proteins.

Chapter 3

Building Your Knowledge

1.
a. Producing a culture by introducing a sample to media.
b. Setting up conditions for bacteria to grow.
c. Separating bacteria into colonies (pure culture).
d. Colonies and cells of bacteria are observed.
e. Determining the species of microbe.
2. The purpose of each technique is to isolate bacteria and develop a pure culture. See figure 3b, d, f.
3. Solid media. The first thing you would do is streak for isolation on a streak plate.
4. Agar is a complex polysaccharide that is used as a solidifying agent for media.
5. Simple media has a known chemical composition. Complex media contains one or more ingredients whose chemistry is unknown. All media with milk or beef extract are complex media.
6. Selective media inhibits the growth of non-target microbes. Differential media allows for the growth of most organisms, but shows visible differences between species. Selective media.
7. MacConkey selects for gram-negative organisms and differentiates between lactose fermentors (pink) and nonfermentors (white).
8. Mixed cultures have multiple KNOWN species. Contaminated cultures have unwanted, UNKNOWN microbes in them.
9. See figure 3.14. a. ocular; b. body; c. nose piece; d. objective lens; e. stage; f. condenser; g. diaphragm; h. base; i. field diaphragm; j. light control; k. stage adjustment; l. fine focus; m. course adjust; n. arm.
10. Magnification is the enlargement of objects so they can be seen. Resolution is the capacity for a system to distinguish two objects as separate objects. The smaller the resolution or resolving power, the clearer the image.
11. Viruses are very small and below the magnifying and resolving power of a light microscope. Therefore, a virologist could not use a light microscope to visualize viral particles.
12. Samples are stained with fluorescent dyes and scanned by a specialized laser to form sections to form a three-dimensional image. This image is very useful when studying intracellular bacteria, since it will give the exact position of the bacterial cell in relation to the host cell.
13. A hanging drop gives more accurate information about the shape, arrangement, and motility of target cells.

14. Gram stains cannot be done to living cells, since the first step of a Gram stain is to heat fix the cells to the slide, which would kill them.
15. A positive stain will add dye to an object, allowing a greater contrast. Simple stains such as methylene blue and differential stains such as a Gram stain are positive stains. Negative stains will cause the background to pick up stain, but the target will remain unstained. A capsule stain is a negative stain.
16. The first step of a Gram stain is applying the sample containing bacterial cells to a slide and gently heating the sample to "fix" the cells to the glass. The next step is apply crystal violet followed by iodine to the sample. Both gram-negative and gram-positive cells will be purple. The alcohol rinse will remove the purple dye from gram-negative cells, but not gram-positive cells. A red safranin counterstain will turn the gram-negative cells from colorless to pink, while keeping the gram-positive cells purple.
17. pink; purple
18. *Mycobacteria* and *Nocardia*
19. a. capsules, b. endospores; c.flagella.

Organizing Your Knowledge

Table 1

a. Semisolid
b. Enriched
c. Thioglycolate added
d. Blood agar
e. Contains buffers, salts, and agents to prevent destruction of sample
f. Carbohydrate fermentation media
g. Contains bile and crystal violet to inhibit gram-positive growth
h. Complex media
i. Isolates *Staphylococcus*

Table 2

a. Dark-field
b. Observation of live, unstained samples
c. Visible light
d. Visualizing internal cellular detail
e. Fluorescent
f. Samples stained with fluorescent dyes
g. Beam of electrons
h. Shows surfaces in great detail as a three dimensional image.

Table 3

a. Wet mount; b. Live; c. Observing arrangement and motility of cells; d. Transmission electron microscopy; e. Used to differentiate between capsulated and non-encapsulated strains of bacteria; f. Use both a primary and a counterstain to differentiate between different types of bacterial cells; g. Spore stain; h. Negative stain; i. Dead; j. Gram stain.

Self-Test: Vocabulary

1. Inoculation	2. colonies
3. spread	4. incubation
5. contaminated	6. selective
7. transport	8. resolving
9. negative	10. complex

Self-Test: Multiple Choice

1. d	4. c	7. d	10. e	13. b
2. c	5. c	8. a	11. b	14. b
3. d	6. d	9. c	12. b	15. e

Applications of Chapter 3

1. Answers will vary, but should include mention that the human body has been extensively sampled in clinical settings—as a means of diagnosing various illnesses.
2. Gelatin is less suitable because it melts at incubation temperatures and because it may be digested by bacteria.
3. Clinical researchers want an accurate estimate of the bacteria present in the sample (therefore in the patient). If some bacteria overgrow others in the transport media, it would be very easy to fail to isolate the actual causative agent.

Chapter 4

Building Your Knowledge

1. Procaryotic and eucaryotic cells.
2. Growth and development, reproduction and heredity, metabolism, movement, cell support, protection and storage, transport systems in and out of a cell.
3. a. fimbriae; b. ribosomes; c. cell wall; d. cell membrane; e. capsule; f. cytoplasmic matrix; g. flagellum; h. inclusion body; i. pilus; j. chromosome; k. actin filaments; l. mesosome; m. slime layer.
4. a. motility; b. attachment.
5. a. figure 4.3d; b. figure 4.3b; c. figure 4.3a; d. figure 4.3c.
6. Attractants inhibit the tumble cycle.
7. Spirochetes have periplasmic flagella. They move by twisting and flexing.
8. Adhesion, conjugation.
9. A slime layer is loosely associated with the cell and protects the cell from nutrient loss and dehydration. A capsule is more tightly bound to the cell and provides adhesion and protection from phagocytosis.
10. Peptidoglycan is a polymer of long chains of sugar molecules joined by short peptide chains. It provides support and protection from lysis due to osmotic pressure in hypotonic solutions. Both lysozyme and penicillin cause disruption of the cell wall by targeting peptidoglycan.
11. In general, gram-negative cells are harder to inhibit or kill.
12. Acid-fast stains identify bacteria with a thick waxy coat (mycolic acid), such as *Mycobacteria* and *Nocardia* species.
13. Mycoplasmas lack a cell wall and use sterols to strengthen their cell membranes.
14. a. selectively permeable barrier and transport functions; b. metabolism (electron transport chain); c. photosynthesis (pigments); d. synthesis of cell walls components.
15. DNA can be found in chromosomes and plasmids in procaryotic cells. The chromosome is larger and carries essential genes. Plasmids are smaller and carry non-essential genes.
16. Ribosomes are required for protein synthesis.
17. An endospore is a dormant form of bacteria that is resistant to most forms of environmental stress. Both *Bacillus* species and *Clostridium* species form endopsores. Endospores are not offspring of vegetative cells, they are a differentiated form. The formation of endospores is not a form or reproduction.
18. a. coccus (round); b. bacillus (rod); c. spirillum (spiral).
19. See figure 4. 25.

20. a. Classification makes identification easier. b. Classification enables scientists to study relationships between bacterial groups.
21. There are thousands of different gram-positive cocci or gram-negative rods and most are not related to one another.
22. rRNA
23. Phenetic classification is based on phenotype or metabolic characteristics; phylogenetic classification is based on evolutionary relationships.
24. Phylogeny.
25. Medical microbiologists use an informal system based on phenotype because it is the phenotype of a particular pathogen that can be easily seen and described rather than its phylogeny.
26. Strains are subspecies of a particular type of bacteria. For example capsulated (S strains) and non-encapsulated (R) strains of the same species of bacteria commonly exist.
27. Both cyanobacteria and the sulfur bacteria are photosynthetic, however, only the cyanobacteria produce oxygen.
28. Myxobacteria are gliding slime bacteria with a very complex life cycle that includes fruiting bodies.
29. a. *Rickettsias*—cause tick-borne diseases such as Rocky Mountain spotted fever and typhus.
b. *Chlamydias* cause trachoma and pneumonia. Neither can be grown on agar plates since they require a host cell.
30. Bacteria and Archaea are procaryotic domains; however, archaea are more closely related to eukaryotes than to bacteria. Archaea have been found in extreme salt, temperature, pressure and pH conditions.

Organizing Your Knowledge

Table 1. 1. d, 2. a, k, 3. c, 4. e, l, i, 5. b, 6. j.
Table 2. a. Adhesion and protection; b. External c. Motility; d. Endospore; e. Internal (cytosol); f. Maintains cell shape; g. Pili or fimbriae; h. internal; i. Storage; j. Ribosomes; k. Cell wall (peptidoglycan); l. External.

Self-Test: Vocabulary

ACROSS: 4 plasmid, 7 archaea, 9 inclusions, 10 procaryotes, 11 purple, 12 glycocalyx, 13 flagella, 14 endospore; DOWN: 1 pili, 2 metabolism, 3 rickettsia, 5 mycoplasma, 6 peptidoglycan, 7 actin, 8 spirochetes.

Self-Test: Multiple Choice

1. a	4. c	7. b	10. a	13. b
2. c	5. c	8. c	11. a	14. d
3. d	6. d	9. b	12. b	15. b

Applications of Chapter 4

Biofilms are a mixed culture of cells that are strongly adherent to the surface they are growing upon. If they are growing on an internal medical device, such as an artificial joint, they are difficult to remove and are nearly impossible to treat with antibiotic therapy. Occasionally biofilms require the removal of the device, which can be a nuisance or require life-threatening surgery.

Chapter 5

Building Your Knowledge

1. Certain organelles (mitochondria and chloroplasts) found in eukaryotes were originally free-living procaryotes that have become endosymbionts (see Insight 5.1).
2. Originally eucaryotic cells were independent that eventually formed colonies. Colonial lifestyle led to specialization of function and structure and the loss of free-living independence in the move to true multicellular forms.
3. a. cell membrane; b. cell wall; c. mitochondria; d. rough endoplasmic reticulum; e. flagellum; f. nuclear membrane; g. nucleolus; h. nucleus; i. centrioles; j. microvilli; k. chloroplast; l. microtubules; m. smooth endoplasmic reticulum; n. microfilaments; o. Golgi apparatus; p. lysosome.
4. Movement toward food or away from noxious stimuli is important for life. Protozoans may use cilia or flagella for movement.
5. Procaryotic flagella are similar in function to flagella (mobility), however they are very different in structure. Eucaryotic flagella are larger, more complex, and have a 9+2 arrangement of microtubules.
6. The glycocalyx is the point of contact between a cell and its environment. It is important in adhesion, response to environmental stimuli and protection.
7. Fungal cell walls are made up of chitin or cellulose and mixed glycans. Algae have cell walls, but their walls are more variable in composition.
8. See figure 5.5.
9. Rough endoplasmic reticulum has ribosomes embedded in it, smooth does not.
10. See figure 5.9.
11. Lysosomes are required for intracellular digestion, phagocytes would have more lysosomes.
12. See figure 5.11.
13. Both chloroplasts and mitochondria have their own DNA, double membranes, and ribosomes.
14. Microfilaments (ameboid motion) and microtubules (move cilia and flagella).
15. When first described, fungi were classified as plants. They are now considered a separate Kingdom in the domain Eukarya.
16. Hyphae are long threadlike cells, yeasts are single-celled organisms, and pseudohyphae are chains of yeasts. Dimorphic fungi have both a yeast and pseudohyphal form.
17. Fungi are heterotrophs, most of which are decomposers (saprobes) that can digest a wide variety of substrates.
18. Most fungi are hardy species that can grow under adverse conditions.
19. Fungus.
20. Mycelium.
21. Septate hyphae have cross-walls; nonsepatate hyphae do not. Vegetative hyphae make up most of the body mass of the fungus, reproductive hyphae produce spores.
22. Both endospores and fungal spores are environmentally resistant forms, however, endosopores are not a form of reproduction and fungal spores are reproductive structures.
23. Asexually produced spores come from a sporangium (sporangiospores) or are free spores derived from the ends of hyphae (condidia).
24. Sexual reproduction, while producing less offspring, will produce more diverse offspring. Zygospores are sexually produced.

25. Zygospores are diploid spores that undergo meiosis to release sporangiospores. Ascospores develop in a sac called an ascus and are haploid. Basidiospores develop in club fungi and are also haploid.
26. Water: potatoes: grapes.
27. Sexual and asexual spore development.
28. Deuteromycetes do not have a known sexual reproductive phase.
29. a. cornmeal; b. blood; c. Sabouraud's agar.
30. Sexual spores are rarely seen in clinical lab culture, so are not generally used to identify clinical isolates.
31. Immunocompromised patients are susceptible to both primary and opportunistic fungal pathogens. Also, fungal spores are nearly everywhere so exposure is common.
32. Fungi are potent decomposers and cause food spoilage, especially in fruits. They may also grow on foods and produce toxins that are harmful if ingested (such as Aspergillus).
33. Algae are plantlike photosynthetic protists, protozoans are animal-like protists that lack cell walls.
34. Algae live in both marine and freshwater aquatic habitats. Some species live in symbiosis with fungi (lichens) and some are desert-adapted.
35. a. Euglenophyta (euglenids); b.Pyrrophyta (dinoflagellates); c. Chrysophyta (diatoms); d. Phaeophyta (brown algae); e. Chlorophyta (green algae).
36. Dinoflagellates may cause red tides, which may cause paralytic shellfish poisoning.
37. Protozoans are animal-like and are heterotrophic.
38. Protozoans move by pseudopod, cilia, or flagella.
39. Trophozoites are the motile, feeding stage. Cysts are a dormant, environmentally resistant stage. Protozoans that do not form cysts would likely not survive the food or water environment so wouldn't be easily transmitted.
40. Mastigophora—including trypanosomes, *Leishmania, Giardia;* Sarcodina—including *Entamoeba;* Ciliophora—Paramecium; Apicomplexa—*Plasmodium.*
41. Most protozoans are identified by microscopy.
42. Most trypanosomes are spread by arthropod bite. *T. cruzi* causes Chaga's disease and *T. brucei* causes sleeping sickness.
43. Amebic dysentery is spread by fecal contamination of food or water. Symptoms include nausea, vomiting, diarrhea, and dehydration.
44. Helminths are multicellular worms. Those of medical importance are parasitic and have well-developed reproductive systems.
45. Microscopic examination of eggs or larvae is sufficient for diagnosis of most helminths.

Organizing Your Knowledge

Table 1. a. Zygospore; b. Ascus containing ascospores; c. Gills with basidia containing basidiospores; d. None found; e. Bread mold; f. *Histoplasma*, ring worm; g. Mushrooms, *Cryptococcus*; e. *Candida.*

Table 2 a. E, b. B, c. E, d. E, e. B, f. E, g. E, h. E, i. B, j. B, k. E, l. E, m. B. 1. Lipid synthesis; 2. Motility; 3. Containment of heredity; 4. Synthesis of proteins; 5. Adhesion and protection; 6. Transport and modification of proteins; 7. Motility; 8. Energy transformation; 9. Protection from lysis; 10. Semi-permeable barrier; 11. Synthesis of ribosomes; 12. Photosynthesis; 13, Protein synthesis.

Self-Test: Multiple Choice

1.	c.	2.	b.	3.	b.
4.	c.	5.	a.	6.	a.
7.	c.	8.	d	9.	c.
10.	b.	11.	d.	12.	c.
13.	d.	14.	a.	15.	a.

Applications of Chapter 5

1. Fungi and humans are both eucaryotic. There are many fewer targets that would harm fungi and leave human cells unharmed than there are antibacterial agents. Chemicals, such as Amphotericin B target structures that both organisms have, so there is not much selective toxicity.
2. Mycoplasmas lack cell walls and have sterols in their membranes. These sterols need to be added to media. Animal cells lack cell walls and have sterols. Plant cells and most bacterial have cell walls to support the cell membrane and lack sterols that are used to support and stabilize the membrane.

Chapter 6

Building Your Knowledge

1. No—viral particles are too small to see.
2. Virus means poison in Latin. Louis Pasteur coined the term as he described the causative agent of rabies.
3. In the late 1890s and early 1900s viruses were discovered that could infect plants (TMV), animals (foot and mouth), and bacteria (bacteriophage). These particles would pass through filters that would exclude larger cells (bacteria). The filtrates were still capable of transmitting disease, indicating the causative agent was much smaller than bacterial cells.
4. Every type of cell tested to date—fungi, algae, bacteria, plants, and animals.
5. Answers will vary. According to the cell theory, viruses aren't alive—they are not made up of cells. Also, scientists have chemically synthesized active viral particles (Insight 6.2). Viruses are not cellular, so are not called organisms.
6. Viruses are not capable of reproducing without a host cell and therefore cannot be grown on nutrient agar (no host cells to support their growth).
Rickettsias are obligate intracellular parasites, like viruses and are very small for a bacterial cell (300 nm). However, *Rickettsias* are prokaryotic cells.
7. See figure 6.4.
8. Nucleic acid protein.
9. Capsomers are self-assembling protein subunits that make up the capsid of a viral particle.
10. Not all viruses are enveloped. The envelope is a lipid coat that was originally part of the host cell's membrane system.
11. Viral spikes are used to attach to a host cell.
12. a. protection from environmental stressors; b. binding to a host cell; c. penetration of viral DNA or RNA into a host cell.
13. DNA or RNA may be found in a viral particle, but not both.
14. Other components MAY (not all viruses have these) include enzymes (reverse transcriptase), ribosomes, or tRNAs.
15. Nucleic acid present (DNA or RNA), envelope presence, single or double strandedness of nucleic acid, and the size and shape of virus.

16. Viral particles bind to (adsorb) specific receptors found on host cells. Some viruses are very specific in both the host and tissue type they will infect, others are more generalized. For example hepatitis B only infects the liver cells of humans, but rabies can infect all mammals.
17. DNA viruses are assembled in the nucleus; RNA viruses are assembled in the cytosol.
18. Enveloped viruses bud out of the host cell and do not lyse the cell with exit. Naked viruses, however, do lyse host cells as the exit.
19. a. Herpes simplex virus and b. herpes zoster virus are both persistent infections. Herpes simplex causes herpes. Herpes zoster causes chickenpox and later shingles.
20. Some viral infections change a host cell's DNA, causing a misregulation of cell growth which causes cancer. Papilloma virus causes cervical cancer, Epstein-Barr virus causes Burkitt's lymphoma.
21. A lytic virus will enter a cell and kill it after using it to make more viral particles. Lysogenic viruses do not immediately lyse the cell, but exist as a prophage indefinitely.
22. 1. adsorption; 2. penetration; 3. duplication of phage; 4. assembly of virions; 5. maturation; 6. lysis of cell.
23. Lysogenic conversion is the change in phenotype due to the incorporation of a prophage. The toxin produced by the organism that causes diphtheria is from a prophage.
24. a. to isolate and identify viruses from clinical samples; b. prepare viruses for vaccines; c. to research viruses.
25. a. cell culture; b. bird embryos (eggs); c. live animal inoculation.
26. Answers vary: influenza, common cold, yellow fever, measles, rabies, HIV, Ebola, SARS, polio, and others.
27. Prions are infectious particles made up of protein alone. This makes them different from viruses, with protein and nucleic acids. Prion diseases include BSE (mad cow disease), variant Creutzfeldt-Jakob, scrapie, and chronic wasting disease.
28. Viriods are RNA-only pathogens that attack plants.
29. Viruses are not bacterial cells, so the antibiotics that target structures on bacterial cells (such as penicillin targeting peptidoglycan or tetracycline targeting bacterial ribosomes) are ineffective.
30. There are almost no unique characteristics that a viral particle would have that the host would not, since the virus "steals" most of the host's metabolic and genetic machinery during the course of the infection.

Organizing Your Knowledge

a. Protection of DNA; b. Outer coat; c. Yes; d. Spikes; e. Protruding from envelope; f. Envelope; g. Protection means of entry; h. No; i. DNA or RNA; j. Interior; k. Yes; l. Change RNA genome to DNA; m. Interior; n. No—retroviruses only.

Self-Test: Vocabulary

ACROSS: 3 adsorption; 5 capsid; 6 envelope; 9 spikes; 11 budding; 15 prion; 17 cytopathic; 18 reverse transcriptase; 19 eggs; 20 uncoating. DOWN: 1 lytic; 2 bacteriophage; 4 rabies; 7 viriod; 8 plaques; 10 prophage; 12 lysogeny; 13 induction; 14 virus; 16 nucleus.

Self-Test: Multiple Choice

1. a.	2. b.	3. a.
4. b.	5. d.	6. c.
7. b.	8. a.	9. c.
10. a.	11. c.	12. c.
13. d.	14. c.	15. b.

Applications of Chapter 6

1. Answers will vary. The culture of viruses is a very time-consuming and expensive process. By the time a definitive diagnosis of the viral cause, the patient will commonly have gotten better without further intervention. In certain cases (such as RSV), the infection is more severe in certain populations and samples will be tested for the presence or absence of viral proteins, which can be done quickly.
2. Answers will vary. Bacteriophage can be used as disinfectants to treat surfaces. They can also be used in food microbiology, adding specific bacteriophage to food to limit the growth and spread of harmful bacteria and in medical microbiology, as topical agents to limit the growth of bacteria.

Chapter 7

Building Your Knowledge

1. Essential nutrients must be acquired from the environment because the organism cannot synthesize these nutrients.
2. a. Relatively large quantities; b. Trace amounts; c. Structures in cell; d. Enzyme cofactors; e. Carbon, hydrogen, oxygen; f. Zinc, cobalt, nickel.
3. Autotrophs = inorganic carbon (carbon dioxide). Heterotrophs = organic sources.
4. Photoautotrophs = energy from sunlight. Chemoautotrophs = energy from chemical bonds.
5. Cyanobacteria and eucaryotic algae get carbon from carbon dioxide and energy from sunlight.
6. Saprobes are decomposers that get their nutrients from dead organisms. Parasites also acquire their nutrients from other organisms; however, parasites use a live host.
7. See figure 7.2. The microbe secretes enzymes outside of its cell wall, breaking down the larger food particles into pieces small enough to cross the cell wall and cell membrane.
8. Several opportunistic pathogens, such as *Pseudomonas aeruginosa* and *Cryptococcus neoformans* are soil-associated saprobes that cause disease in immunocompromised hosts.
9. Obligate intracellular parasites require host cells to be cultivated in lab, so cannot be grown by themselves on agar plates.
10. All viruses, rickettsias, chlamydias, and the parasite that causes malaria are obligate intracellular parasites.
11. Molecules will spontaneously move from areas of high concentration to areas of low concentration by simple diffusion.
12. See figure 7.5. Cells in hypertonic solutions shrink, those in hypotonic solutions swell. The cell wall protects against hypotonic lysis.
13. Facilitated diffusion requires a carrier protein to transport molecules across a membrane. Like simple diffusion, it does not require energy, however the carrier

proteins can be saturated which would limit the rate of entry into a cell.
14. Active transport is required when moving molecules against a concentration gradient (from low to high), moving molecules faster than simple diffusion, or if the molecules will be chemically modified. Substrates entering a cell by group translocation are chemically modified, readying them for metabolism.
15. a. minimum; b. optimum; c. maximum.
16. Microbes vary greatly in their survivable temperature ranges. Psychrophiles live in the cold, mesophiles in moderate temperatures and thermophiles live at temperatures near the boiling point of water. Facultative psychrophiles are of concern to food microbiologists because they can grow at refrigeration temperatures as well as the human body. Most human pathogens are mesophiles.
17. These organisms differ on their oxygen requirements. Obligate aerobes require oxygen, microaerophiles require small amounts of oxygen, facultative anaerobes can live with or without oxygen and obligate anaerobes are poisoned by oxygen.
18. See figure 7.11.
19. Salt (osmotic pressure).
20. Pressure.
21. In satellitism one organism is unaffected by the growth of a second organism, but the second organism requires nutrition or protection from the first organism.
22. The nitrogen fixer converts atmospheric nitrogen to an ammonia compound that is usable by the cellulose digesting microbe. The second microbe degrades cellulose, making the carbon available to the nitrogen fixer.
23. In antagonism, one microbe thrives by attacking other organisms in the community. Microbes that produce antibiotics are antagonists.
24. Biofilms are organized communities of microbes attached to a substrate and each other by extracellular matrix. Bacteria in a biofilm are able to communicate with one another by quorum sensing—mechanisms to determine cell density and response to environmental stimuli.
25. Biofilm bacteria are very resistant to antibiotics and may form a biofilm on medical implants.
26. Bacteria reproduce by binary fission. See figure 7.14.
27. Generation time varies greatly. *Mycobacterium* spp. are very slow-growing (doubling in 10–30 days), most pathogens have generation times under an hour.
28. See figure 7.15.
29. Exponential.
30. Nutrients run out and toxins build up in the media, inhibiting growth.
31. 5 hours = 10 doublings = 2^10*25 = 25,600 CFU.
32. a. turbidity; b. direct microscopic count; c. Coulter counter; d. real-time PCR; e. colony-forming units (plate counts). Plate counts will ONLY count live cells.

Organizing Your Knowledge

Table 1

a. Carbon dioxide; b. Organic sources; c. Proteins (amino acids) nucleic acids, peptidoglycan; d. Nitrogen fixers use atmospheric nitrogen (N^2), all other organisms use some form of ammonia; e. Terminal electron acceptor for aerobes, major component to macromolecules (carbohydrates especially); f. Phospholipids (membranes), nucleic acids (including ATP); g. Sulfur salts (mineral deposits) H_2S.

Table 2

a. Cold (below 15° C); b. Acidophile; c. No atmospheric oxygen; d. Microaerophile; e. Basic pH; f. Halophile; g. moderate temperatures (20–40° C); h. Hyperthermophiles (between 80–25° C); i. Requires oxygen; j. Barophile.

Table 3

a. +/+, b. +/0, c. +/-, d. +/+, e. +/-

Table 4

a. Cells are preparing to divide—making ribosomes; enzymes and substrates for growth; b. Slow (non-growing), c. Fast; d. Nutrients are running out, toxins are building up; e. Dead cells are accumulating; f. Dead cells begin to outnumber the live cells.

Self-Test : Vocabulary

1. S	2. U	3. P
4. Y	5. N	6. V
7. I	8. J	9. Q
10. W	11. O	12. C
13. Z	14. D	15. H

Self-Test: Multiple Choice

1. c	2. d	3. d
4. b	5. d	6. d
7. b	8. b	9. c
10. b	11. a	12. b
13. b	14. d	15. a

Applications of Chapter 7

1. Obligate psychrophiles cannot grow at body temperatures, facultative psychrophiles can grow in food in the refrigerator, and in the human body.
2. Answers will vary. Potential targets include quorum-sensing proteins or any components of the quorum-sensing pathway—you could "trick" the system into sensing more bacteria were present, thereby inhibiting bacterial growth. Also, using enzymes that degrade carbohydrates, to break apart the matrix of the biofilm.

Chapter 8

Building Your Knowledge

1. a. Biosynthesis, bond making; b. Breaking down larger molecules; c. Requires energy; d. Releases energy.
2. A. & B. Carbohydrates, lipids, proteins; C. Sugars, amino acids; D. ATP; E. Releases energy; F. Requires energy.
3. Catalysts speed chemical reactions without adding energy to a reaction or being changed by the reaction they catalyze.
4. Most enzymes will catalyze a reaction for many substrate molecules, so a small addition of enzyme will increase the product produced dramatically.
5. Proteins.
6. Enzymes speed reactions by lowering the energy of activation of the reaction.
7. Holoenzymes are the complete, functional enzyme; apoenzymes are the catalytic protein, in the absence of its cofactors (either proteins or metallic ions).
8. Enzymes react with a limited number of substrates—like a lock will only work with a specific key or limited similar keys. However, the active site isn't a static platform—the substrate interaction causes changes to its structure (induced fit).

9. Exoenzymes.
10. Constitutive; substrate.
11. Condensation (dehydration).
12. Breaks (hydrolysis).
13. Oxidation is the loss of electrons; reduction is the gaining of electrons. When one substrate in a chemical reaction is oxidized, another is reduced, conserving the total number electrons present.
14. a. Streptokinase (dissolves clots); b. Elastase and collagenase (dissolves connective tissue); c. Lecithinase C (damages cell membranes); d. Strep throat; e. Lung and burn infections; f. Gas gangrene.
15. Low temperatures, suboptimum pH or substrate concentration will lower enzyme activity. Enzyme activity will be destroyed at high temperatures as enzymes denature. Enzymes may also be oxidized or exposed to heavy metals and be permanently inactivated.
16. Most metabolic reactions are steps in a metabolic pathway, with products of one reaction being the substrates of one or more other enzyme-catalyzed reactions (see figure 8.10).
17. Enzyme *activity* may be altered in two different ways. Competitive inhibitors, that bind to the active site and noncompetitive inhibitors that bind to the regulatory site of an enzyme. The *amount* of enzyme available may be altered by either repressing or inducing the production of proteins at a genetic level.
18. Energy is the ability to do work. Different forms of energy include heat, wave, electrical, mechanical, atomic or chemical. Chemical energy is most commonly used in cells.
19. Exergonic reactions release energy and are typically catabolic; however, endergonic reactions require energy and are generally anabolic.
20. Glucose is a reduced form—during metabolism (glycolysis and TCA cycle) the bonds of glucose are oxidized and the electrons are taken by electron carriers to the electron transport chain.
21. ATP is adenosine triphosphate and can be stored, spent, exchanged, and earned by cellular processes, making the "metabolic money" analogy an accurate one. A. High energy phosphate bond; B. Ribose; C. Adenine.
22. Aerobic respiration uses oxygen as the final electron acceptor at the end of the electron transport chain and produces more energy than anaerobic respiration.
23. 6, 6, 6, 38.
24. Oxygen.
25. Oxidize.
26. Glucose, two pyruvate, 2 ATP (also 2 NADH that may go to the electron transport chain in respiring organisms).
27. 6,3, 2.
28. Carbon dioxide, ATP, electrons, and loaded carriers (NADH, FADH2) that go to the electron transport chain with their electrons.
29. Electron transport chain, there must be a flow of protons (H+) through the ATPase molecule to catalyze the addition of a phosphate to ADP.
30. The flow of protons causes a shape change in the ATPase molecule, which catalyzes the reaction.
31. See figure 8.24.
32. Pyruvic acid may go to the TCA cycle or may serve as an electron acceptor for NADH produced in glycolysis (fermentation reactions). Pyruvate can become ethanol, various acids, acetone and several other molecules as fermentation continues (see figure 8.20).
33. Most ATP is generated in the electron transport chain.
34. Fermentation uses pyruvate as an electron acceptor for NADH and produces much less energy than anaerobic respiration.
35. Alcoholic fermentations produce ethanol and CO^2. Acidic fermentations produce various acids (different with different organisms). Beer or bread making rely on alcoholic fermentation while milk sours or changes to yogurt as a result of acidic fermentation.
36. Carbon for amino acid synthesis comes from glycolysis and pyruvate. Amino acids combine to form proteins.
37. Carbohydrates are produced by gluconeogenesis, the building rather than breaking down of sugar molecules. These molecules are used for cell walls and as storage.
38. Acyl carrier protein takes acetyl groups from acetyl CoA and adds them to a chain of carbon. Lipids are used as membrane components and for storage.
39. Amphibolism means a pathway can be both anabolic (building) or catabolic (tearing down).
40. No—the other parts of metabolism may provide precursors, the electron transport chain does not.
41. Carbon from sugar would exit the pathway as carbon dioxide, in the absence of any anabolic reactions.

Organizing Your Knowledge

Table 1. a. Pyruvic acid; b. Pyruvic acid, NAD+; c. Carbon dioxide, NADH, FADH2; d. Membrane.
Table 2. See table 8.3 in text.

Self-Test: Vocabulary

1. G	2. H	3. P
4. Y	5. J	6. R
7. L	8. V	9. T
10. M	11. U	12. O
13. E	14. S	15. W
16. N	17. C	18. F
19. I	20. D	

Self-Test: Multiple Choice

1. b	2. a	3. d
4. b	5. b	6. b
7. c	8. d	9. c
10. b	11. b	12. d
13. d	14. c	15. d

Applications of Chapter 8

1. Answers will vary. Ancient Earth did not have free oxygen, so the first bacteria were very likely anaerobes that produced energy through fermentation or anaerobic respiration.
2. Answers will vary. Microbes used for bioremediation cannot be poisoned by the pollutant they are intended to degrade. In a perfect situation, they should use this pollutant as a carbon source rather than glucose and be anaerobes that produce enough energy to thrive in the soil as they degrade the pollutant.

Chapter 9

Building Your Knowledge

1. a. transmission of traits (genes) from parent to offspring; b. expression and variation of traits; c. structure and function of genetic material; d. how genetic material changes.

2. A genome is the total of all genetic material in an organism. Genomes may be arranged as several linear chromosomes (most eucaryotes), a single circular chromosome (most procaryotes), chromosomal and plasmid DNA (found in most procaryotes), and DNA or RNA (viruses).
3. The definition of a gene varies—classical genetics describes a gene as the basic unit of heredity responsible for a trait, molecularly it is the site on a chromosome that provides instructions for a single cellular function or, most basically, it is the site on a chromosome that codes for a protein or RNA molecule.
4. Structural genes, regulatory genes, RNA.
5. Much longer (1 mm vs. 1 um in *E. coli*).
6. Phosphate, deoxyribose, nitrogenous base. For drawing, see figure 9.4.
7. Complementary pairs are made from nitrogenous bases that bind together with hydrogen bonds, forming the "rungs" of the DNA ladder. Purines bind with pyrimidines, specifically A binds to T and G binds to C in DNA, in RNA U replaces T.
8. Covalent sugar-phosphate bonds which are much stronger than the hydrogen bonds that hold the base pairs together.
9. See figure 9.4.
10. The strands run 3′–5′ on one strand and the opposite strand will run 5′-3′.
11. a. Maintenance of genetic code since each strand is the template for a new strand during replication. b. Options for variety and storage of information.
12. a. Primase; b. Gyrase; c. Unwinds helix to start replication; d. Seals nicks in backbone and joins fragments.
13. More mutations would occur, since DNA polymerase I detects and repairs mismatched bases.
14. See figure 9.6.
15. RNA polymerase does NOT require a 3′ OH to produce RNA from DNA, which makes it valuable in replication, since it can synthesize a primer made of RNA that provides a 3′OH for the DNA polymerase III to add on to.
16. DNA is antiparallel and DNA polymerase III can only add to the 3′ end, so one strand (the lagging strand) is primed many times and made up of short segments.
17. See figure 9.8.
18. DNA sequence determines protein sequence, which determines the shape and function of the protein. The shape and functionality of a protein determine the phenotype of an organism. Changes in DNA sequence may lead to changes in amino acid sequence, which may lead to changes in the structure and function of a protein.
19. a. RNA is single-stranded and has variable secondary and tertiary structures. b. RNA has uracil instead of thymine binding to adenine. c. RNA has ribose instead of deoxyribose.
20. See figure 9.11. Initiation—RNA polymerase binds to DNA at the promoter. Elongation—ribonucleotides are added to the 3′end of the growing chain. Termination—RNA polymerase releases DNA.
21. a. tRNA; b. rRNA; c. Message carrier from DNA; d. Structural component of ribosome; e. No; f. No.
22. See figure 9.11.
23. a. initiation; b. elongation; c. termination.
24. See figure 9.1. It is a triplet of mRNA that codes for an amino acid.
25. The same codons code for the same amino acids in nearly all organisms. There are mitochondrial genes that place an amino acid in the place of what is most commonly a stop codon.
26. See figure 9.12.
27. Met Leu Leu Pro Ala Stop.
28. A promoter is the region of DNA recognized by RNA polymerase.
29. A nonsense codon will terminate translation.
30. A polyribosomal complex is a segment of mRNA with multiple ribosomes attached. Polyribosomes do not exist in eucaryotes.
31. Introns are sequences in the middle of genes that are transcribed, but are not translated because they are removed during RNA processing. These are found in eucaryotes, but not procaryotes. Exons are regions that code for protein.
32. DNA viruses are generally double-stranded, RNA viruses are generally single-stranded. DNA viruses are usually replicated in the nucleus, using the host cell's DNA polymerase.
33. a. See figure 9.18a, b. See figure 9.18b.
34. Corepressors bind to repressors to stop transcription.
35. Examples of antibiotic targets include: RNA polymerase (rifamycins), mRNA elongation (actinomycin D), ribosome attachment to mRNA (erythromycin), polypeptide elongation (tetracycline).
36. Wild type are those strains that are normally found in nature. Mutant strains are those with changes in their DNA and likely phenotype.
37. The Ames test is a simple method to test the carcinogenic potential of chemicals by testing their ability to induce mutations. The procedure is described in figure 9.21. Mutagens containing samples will show more colonies than nonmutagens.
38. Spontaneous mutations have no known cause. Induced mutations happen after exposure to mutagens. Organisms vary greatly in their spontaneous mutation rate.
39. Both ethidium bromide and nitrogenous base analogs are chemical agents that induce mutations. EtBr intercalates between the bases of DNA and causes small insertions and deletions. Base analogs induce mispairing. Xrays are physical mutagens and cause breaks in the sugar–phosphate backbone.
40. All three are point mutations. A same sense mutation changes the DNA, but the amino acid remains the same. A missense mutation changes the amino acid sequence and a nonsense mutation introduces a stop codon.
41. Both mutations affect all the amino acids downstream of the mutation as well as the site of the changed DNA and are generally more severe.
42. One bacterial cell donates DNA to another cell, either directly (conjugation) or indirectly (transformation, transduction).
43. a. Transfer of naked DNA from the environment to a cell; b. Conjugation; c. Viral-mediated transfer.
44. See figure 9.22.
45. In generalized transduction, lysis occurs with random host DNA being incorporated into phage

particles. In specialized transduction, a viral prophage is excised from the host genome with host DNA included.
46. A transposon can be excised directly and move, replicate itself, then move, or move onto a plasmid and move to another cell.

Organizing Your Knowledge

Table 1

a. Pyrimidine; b. Purine; c. Adenine; d. Cytosine; e. Purine; f. Pyrimidine.

Table 2

a. RNA polymerase; b. Reverse transcriptase; c. Ribosome; d. DNA polymerase; e. Transcription; f. Reverse transcription; g. Translation; h. Replication.

Table 3

a. Unwinds DNA; b. Makes primers; c. Synthesizes DNA; d. Supercoiling; e. triplets that code for an amino acid; f. Genes that are regulated together; g. Ribosomal RNA; h. Replication; i. Replication; j. Replication; k. Transcription; l. Transcription and translation; m. Translation; n. Translation.

Self-Test: Vocabulary

ACROSS: 1 point, 4 transfer, 8 anticodon, 9 genome, 10 frameshift, 11 transcription, 14 induced, 18 specialized, 20 primase, 21 ligase, 22 introns, 23 polyribosome; DOWN: 2 transposons, 3 gene, 4 transformation, 5 promoter, 6 semiconservative, 7 ames, 12 operon, 13 nitrogenous, 15 redundancy, 16 ribose, 17 phenotype, 19 purines.

Self-Test: Multiple Choice

1.	a	2.	c	3.	d
4.	c	5.	b	6.	a
7.	d	8.	d	9.	c
10.	c	11.	d	12.	d
13.	b	14.	b	15.	c

Applications of Chapter 9

1. The enzymes found in any animal's liver will generally chemically modify compounds to "detoxify them. " There are some mutagens that are not harmful in their native state, but if they were modified by liver enzymes in vivo, they would be very mutagenic.
2. Genetic recombination leads to transfer of genes between organisms in bacteria. Once an organism has developed antibiotic resistance, it may be able to pass on this trait by transformation, transduction, or conjugation.

Chapter 10

Building Your Knowledge

1. Genetic engineering is the direct, deliberate manipulation of an organism's genome.
2. a. selection for specific, desirable traits in plants and animals; b. repair of genetic disorders.
3. Restriction enzymes cut DNA at specific palindromic sequences, that read the same 3′–5′ and 5′–3′. Sticky ends enable scientists to easily splice DNA from one source to another.
4. Ligase is an enzyme that seals nicks in the sugar–phosphate backbone that remain after DNA from two sources is spliced together. *In vivo*, ligase will seal the nicks that remain between Okazaki fragments.
5. Complementary DNA (cDNA) is made by reverse transcribing mRNA to DNA with a specific enzyme (reverse transcriptase). cDNA does not have introns.
6. See figure 10.2. Gel electrophoresis subjects samples to an electric field and will separate DNA based on size. DNA fragments, with their negative charge, will move toward the positive charge. Larger pieces move slower and are retained at the top of the gel. Smaller pieces move faster and are at the bottom.
7. Gene probes are small fragments of DNA that are labeled, usually with a radioactive or fluorescent marker. The probe matches with its complement either on a membrane (Southern blots and library screening) or other medium (microarrays).
8. See figure 10.5. Dideoxynucleotides lack a 3′OH to add new nucleotides onto, so they terminate the chain.
9. Template DNA, primers, nucleotides, buffer, and Taq polymerase are needed to complete a PCR reaction.
10. The heating step needed to denature the DNA will denature most enzymes. While DNA will reanneal and be functional, a heat-denatured enzyme is nonfunctional. A heat-stable enzyme is needed for PCR.
11. Recombinant DNA is DNA from two different sources that is combined.
12. See figure 10.9. If the recombinant plasmid confers antibiotic resistance, only those cells that have the plasmid will grow on media with that antibiotic.
13. A cloning vector must have: a. an origin of replication, b. markers (resistances genes or enzyme function), and c. the ability to accept DNA.
14. See table 10.1. Cloning hosts need to grow fast, with easy culture techniques, should secrete a high yield of protein, be mapped or sequenced and not be pathogenenic.
15. Recombinant proteins are easy to produce, will not be rejected as foreign (animal proteins are often rejected), and have a minimal risk of disease transmission (human fluid transfer has been the source of infections). See table 10.2 for a partial list of recombinant proteins used in human medicine.
16. Transgenic organisms are recombinant organisms that carry genes from other sources.
17. See table 10.3.
18. See table 10.4.
19. Gene therapy adds a gene to an individual. It differs from recombinant protein production in that the host (patient) will then make their own protein rather than needing to be supplemented all the time.
20. See figure 10.12. Uses of DNA fingerprinting include forensics (positively identifying individuals), determining parentage (pedigrees for animals, custody cases), identifying individuals that cannot be identified by other means (due to disaster or damage to a corpse), identifying microorganisms, and detecting genes.
21. Most forensic samples contain very little DNA (trace amounts) the amplification possible with PCR makes these small amounts usable for fingerprinting. (e.g., scientists can identify the person who licked the back of a stamp from the DNA present in the saliva).
22. Microarrays track the expression of genes in individuals, or under specific conditions.

Organizing Your Knowledge

Table 1. Nature vs. Lab

a. Joining Okazaki fragments; b. Cutting DNA at specific sites; c. Making cDNA from mRNA.

Table 2. Methods of Genetic Engineering

a. Determine the order of nucleotides in DNA; b. Determine the nature of DNA present in a sample—both size and binding to probe; c. Multiple answers—may provide source of proteins (biopharming) and research opportunities; d. Amplification of DNA; e. Analysis of the genes expressed by an organism, in certain conditions; f. Analysis of DNA to identify individuals; g. Production of recombinant organisms; h. Identifying samples that will bind to a probe (hybridize).

Self-Test: Vocabulary

1.	A	2.	G	3.	P
4.	K	5.	I	6.	Q
7.	E	8.	F	9.	H
10.	D	11.	M	12.	N
13.	B	14.	T	15.	S
16.	L	17.	C	18.	J

Self-Test: Multiple Choice

1.	c	2.	d	3.	b
4.	c	5.	b	6.	d
7.	d	8.	c	9.	a
10.	b	11.	d	12.	a
13.	d	14.	c	15.	c

Applications of Chapter 10

1. Answers will vary. Altering the eggs and sperm would affect not just the current generation, but also generations in the future. It will also alter an individual's reproductive capacity.
2. Answers will vary. Some genetic defects are dominant disorders (Huntington's disease for example). Adding more of the right gene would not help if the wrong gene product will affect phenotype in a dominant fashion. Also—most genetic disorders are not from a single gene and even those that are a single gene are difficult to supplement the right gene, in the right concentration and with the correct expression to correct a disordered phenotype.

Chapter 11

Building Your Knowledge

1. Contaminants are microbes that are present at a place or time that makes them unwanted.
2. Decontamination methods are generally means of physical control, chemical control, or the mechanical removal of microbes.
3. Physical control methods will change the environment enough to kill microbes—examples include heat treatment and radiation. Chemical control methods will damage microbes by adding chemical agents to an object or solution.
4. Contaminants vary greatly in their resistance to physical and chemical control methods. Endospores and prions are the most resistant; vegetative cells and enveloped viruses are the least resistant.
5. Disinfection removes growing (vegetative) disease-causing agents but not endosopores. Sterizilization removes ALL organisms—including endosopores and viruses. Objects are either sterile or not, there is no "almost sterile. "
6. Anything with a -cidal ending will kill, -static agents simply stop growth. If you take a bacteriostatic antibiotic, if your immune system hasn't cleared the infection, you will get the symptoms back when you stop taking the antibiotics.
7. a. Inanimate objects; b. Living tissue; c. Bleach, boiling; d. Alcohol, hydrogen peroxide.
8. Sanitization is the removal of debris to reduce the potential for infection—it is not the same as sterilization.
9. A microbe is dead if it no longer grows (permanently) in media and conditions known to support its growth.
10. Vegetative cells are easier to kill and have a more rapid death rate than endosopores. Also, the larger the population, the slower the death rate.
11. Higher concentrations of chemical agents will generally kill faster than lower concentrations (notable exception—ethanol works better at 70% than at 100%).
12. a. cell wall, b. cell membrane, c. protein synthesis processes, d. proteins.
13. Detergents, alcohols, and some antibiotics (e.g., penicillin) target the cell wall.
14. Surfactants target the cell membrane by lowering the surface tension of the membrane, causing disruption and leakage.
15. See figure 11.4.
16. a. Moist heat exposure denatures proteins. b. Dry heat dehydrates and will oxidize proteins. Moist heat works at lower temperatures and shorter exposure times.
17. Organism B is more resistant (takes longer to kill).
18. An autoclave uses steam under pressure to reach temperatures high enough to kill bacterial endosopores in a short amount of time (121° C for 20 minutes). The pressure will increase the temperature of the steam ($PV = nRT$).
19. Intermittent steam is used to allow endosopores time to germinate into vegetative cells that are killed at lower temperatures.
20. Boiling water will disinfect the water (and objects in the water). It will not kill endosopores, however.
21. Pasteurization heats foods and liquids to kill infectious and spoilage agents. It does not kill all organisms, so does not sterilize.
22. Cold slows growth of microbes, but does not kill them. In fact, cold will preserve cultures (lyophilization).
23. Ionizing radiation breaks the sugar–phosphate backbone of DNA, killing microbes. Nonionizing radiation introduces thymine dimers which causes mutations to accumulate, killing the microbes.
24. Nonionizing radiation will not penetrate barriers.
25. Filtration can remove contaminants from both liquids and the air. It will not alter the taste of the liquids being filtered (milk, beer, water).
26. Aqueous solutions are water-based, and tinctures have solutes dissolved in alcohol or alcohol–water mixtures.
27. Answers will vary. Rapid action, used in low concentrations, nontoxic, cheap, broad-spectrum activity, noncorrosive, persistent action.
28. Bleach acts by denaturing proteins (oxidizes S–S).
29. Alcohols disrupt membrane function and coagulate proteins. Pure alcohol is less effective because water is needed to coagulate proteins.
30. Detergents disrupt cell membranes (they are amphipathic and break up the phospholipid bilayer).
31. Heavy metals bind to proteins, specifically enzymes, and irreversibly inactive them.
32. See figure 11.17. Glutaraldehyde cross-links proteins.
33. Ethylene oxide is toxic (carcinogenic) and explosive. It is also a gas, not a liquid.

Organizing Your Knowledge

a. Physical; b. Physical; c. Physical; d. Chemical; e. Chemical; f. Chemical; g. DNA and proteins (replication); h. Whole cell—excludes microbes; i. Cell membrane; j. Inactivates enzymes and other proteins; k. Removes debris; l. Metabolic process—slows growth; m. Incinerator; n. Chlorine (swimming pools); o. UV radiation; p. Mild disinfectants (eye solutions); q. Dishwashing.

Self-Test: Vocabulary

ACROSS: 1 sterilization; 5 denature; 7 DNA; 9 bacteriocide; 10 autoclave; 12 lyophilization; 13 pasteurization; 15 ionizing; 16 antiseptics; 18 fungistatic; 19 dessicated; 20 halogens. DOWN: 2 tyndallization; 3 membrane; 4 surfactants; 6 thermoduric; 8 endospore; 11 sanitization; 14 asepsis; 17 phenol.

Self-Test: Multiple Choice

1.	a	2.	b	3.	d
4.	b	5.	b	6.	c
7.	a	8.	b	9.	c
10.	c	11.	a	12.	c
13.	d	14.	b	15.	a

Applications of Chapter 11

1. Answers will vary greatly but should describe the need for sterility (internal implant) and lack of autoclave, so no incinerating or other heat-related methods. Typical responses would include glutaraldehyde, ethylene oxide, ionizing radiation as sterilants.

Chapter 12

Building Your Knowledge

1. One-third.
2. See table 12.1. The perfect antibiotic does not exist because there are often trade-offs. For example, Colistin is very effective against gram-negative bacteria, but is toxic to vertebrate kidneys and there are frequent allergies that develop to it, so it is not widely used.
3. Most antimicrobials are the natural products of bacteria and fungi, which use these compounds to inhibit other microbes in their environment.
4. See table 12.2. The treatment of Strep throat is a chemotherapeutic use of antimicrobials. Taking penicillin every day "just in case" would be prophylactic.
5. Antimicrobial agents are designed to be selective in their toxic effects. They are toxic to target organisms (microbes) but safe (or less toxic) to the vertebrate host.
6. Drugs that target processes or structures not found in the vertebrate host are commonly the most selectively toxic. Examples would be bacterial cell walls or flagella. Drugs that target structures found in both are least selective in their toxicity—membrane disruptors are not taken internally because they are so toxic.
7. The cell wall inhibitor would be more selectively toxic than the plasma membrane inhibitor—both host and microbe have membranes, but only the bacterial pathogen has a cell wall.
8. Broad-spectrum drugs target structures or processes shared by many different microbes, while narrow-spectrum antibiotics will target structures that are unique to a smaller number of microbes.
9. Penicillin will only work against actively growing cells because they inhibit cell wall synthesis.
10. Membrane disruptors are taken externally because they are toxic to both bacterial and host cell membranes.
11. Both chloroquine and AZT block DNA synthesis. However, chloroquine cross-links DNA, and AZT is a nucleotide analog that inhibits viral replication.
12. Procaryote ribosomes are different in size and structure. However, some antimicrobials that target ribosomes can harm mitochondrial ribosomes.
13. See figure 12.6. Sulfonamide binds to an enzyme in the folic acid pathway as a competitive inhibitor—binding to the enzyme, but not being changed by it so it's "used up. " Mammals don't synthesize folic acid, so are unaffected.
14. See figure 12.2.
15. Most penicillin is produced by purification of microbial fermentation.
16. See table 12.5.
17. Semisynthetic penicillins have a broader spectrum, especially against gram-negatives.
18. Allergies and antibiotic resistance limit penicillin's usefulness.
19. Clavulanic acid inhibits beta-lactamase enzymes, so reduces the penicillin resistance.
20. Cephalosporins are beta-lactam antibiotics that inhibit cell wall synthesis. They are poorly absorbed through the intestine and are most commonly given intravenously. Each generation of cephalosporin is more effective against gram-negative bacteria and therefore more broad-spectrum than the last.
21. Fluoroquinolones such as ciprofloxacin are synthetic antibiotics that inhibit DNA synthesis in bacterial cells.
22. Aminoglycosides are antibiotics from actinomycetes that inhibit protein synthesis and are useful to treat gram-negative infections.
23. Tetracyclines bind to ribosomes and stop protein synthesis. They are not commonly used because they cause stomach upset, teeth discoloration, and may cause problems with fetal bone development.
24. Chloramphenicol is toxic to humans and may cause bone marrow damage leading to aplastic anemia.
25. These are macrolide antibiotics that attach to the 50S ribosome subunit of the procaryotic ribosome. Clindamycin tends to cause serious problems in the gastrointestinal tract.
26. These drugs aren't found in nature, so the resistance genes for them should not be naturally selected for.
27. Sulfa drugs are purely synthetic—originally from aniline dyes. This is different from most antibiotics like penicillin, which was purified from a natural source.
28. Fosomycin and synercid are both newly developed antibiotics that aren't widely used because scientists do not want resistance to rapidly develop.
29. Fungi are eucaryotes—as are humans so there is less selective toxicity.
30. See table 22.3. Macrolide polyenes can treat systemic infections (Amphotericin B) or topically (nystatin). Azoles are used to treat both systemic and skin infections. Polyenes are effective against fungal membranes because fungi have ergosterol, which mammalian cells lack.
31. It is difficult to selectively kill parasites without killing the host as well. Malaria is commonly treated with quinine, roundworm infections are treated with mebendazole and thiabendazole.

32. Viruses have very little in their particles—most of their cycle uses host organelles, so there is very little to selectively target that wouldn't kill the host as well.
33. Mumps and measles are rare because we vaccinate against them.
34. a. inhibit entry; b. block nucleic acid synthesis; c. inhibit release or maturation.
35. HIV has reverse transcriptase, which human cells do not produce and can be targeted with minimal host cell damage.
36. See table 12.6.
37. Chromosomal resistance is generally a spontaneous random mutation that does not spread in a population.
38. R factors are plasmids that carry resistance genes. They can spread through multiple populations (see figure 12.13).
39. See figure 12.14.
40. Beta-lactamases are enzymes that break apart the Beta-lactam ring in penicillins and cephalosporins.
41. Pumps, often called MDR (multidrug resistance) pumps, give resistance to a wide range of drugs and even detergents.
42. Sulfa drug resistance can be acquired by forming alternate metabolic paths to produce folic acid.
43. See figure 12.15. Exposure to antibiotic kills susceptible cells, leaving only resistant cells that then multiply because they lack competition.
44. It is very rare that a bacterial group will be resistant to multiple antibiotics (in the absence of MDR pumps).
45. a. toxicity to organs; b. allergic responses; c. disruption of normal flora.
46. Most drugs are metabolized and cleared by the liver or the kidney, so that is often where damage can occur.
47. Tetracylines can cause bone development problems.
48. Some antibiotics directly irritate the intestine, all antibiotics disrupt the normal flora in the gut.
49. No, penicillin allergies often take time to develop. The first dose could be a "sensitizing" dose.
50. a. the nature of the microbe; b. the sensitivity of the microbe; c. medical condition of the patient.
51. Kirby-Bauer and tube dilution tests are two methods to determine the MIC.
52. See figure 12. 18. The larger zone (drug A) is more effective.
53. The MIC is the minimum inhibitory concentration and it is used to determine the relative effectiveness of various antibiotics against a given bacterial population.
54. The therapeutic index is the ratio of the dose of the drug that is toxic to humans to the MIC of the drug. The lower the TI, the more dangerous the drug, so a TI of 0.5 would be very dangerous (and not given, because its TI is less than one—it would kill the host before the microbes).
55. Physicians need to consider the history of the patient (pregnancy?), any underlying medical issues (any atypical absorption of drugs would be a problem) and overall health of the patient as well as the microbe and its resistance to any antibiotics of choice.

Organizing Your Knowledge

Table 1.

a. Inhibits folic acid synthesis; b. Blocks protein synthesis; c. Disrupts fungal membranes (ergosterol); d. Causes build-up of toxins; e. Disrupts fungal membranes; f. Binds to ribosomes; g. Broad spectrum; h. Malaria, i. Skin infections, prophylaxis; j. Tuberculosis; k. HIV, l. Rickettsias, chlamydia.

Self-Test: Vocabulary

1. E	2. O	3. C
4. S	5. R	6. L
7. I	8. N	9. G
10. B	11. V	12. D
13. M	14. P	15. A
16. Q	17. K	18. T
19. J	20. F	

Self-Test: Multiple Choice

1. d	2. b	3. d
4. a	5. b	6. b
7. b	8. d	9. c
10. c	11. c	12. a
13. c	14. a	15. c

Applications of Chapter 12

1. Answers will vary but should include consideration of the time frame involved in identifying not only the pathogen responsible for the disease, but its sensitivity. In a hospital setting this may take 2–4 days; during this time patients are often put on broad-spectrum antibiotics to increase the chances of coverage. The alternative is to not treat the patient at all, or use a potentially incorrect narrow-spectrum antibiotic.
2. Answers will vary. Completely synthetic drugs should not have many intrinsically resistant organisms. Also, there should be less chance of developing, then spreading resistance. However, since these synthetic formulas haven't been selected for, they may not be as effective.

Chapter 13

Building Your Knowledge

1. Pathogens are able to invade or penetrate into normally sterile tissues and cause disease.
2. See figure 13.1. After contact, microbes may colonize without invasion (normal flora), they may penetrate into tissues and multiply in normally sterile tissues (infection). Once infection has occurred microbial action or even immune responses may cause damage to host tissues, followed by recovery and immunity or death.
3. After disease develops the host may hold the pathogen's growth and/or clear the pathogen. In this case, recovery and immunity to this pathogen will follow. A disease state may also progress to death or long-term morbidity.
4. See table 13.1. Generally any surface that has contact with the outside is a habitat for normal flora. Transient flora are present, but not for the long term and do not colonize (reproduce in or on a host). Resident flora reproduce on a host.
5. See table 13.2. Generally any location that is internal and has no contact with the outside environment is normally sterile.
6. Normal flora may change the environment (e.g., pH of vagina) to inhibit pathogen growth; they also may simply take up enough space that pathogens are unable to colonize. There is also evidence of direct antagonism, with resident microbes actively inhibiting other microbes.
7. Endogenous infections are acquired from the overgrowth of normal flora. Immunocompromised individuals are at more risk for these types of infections.

8. See figure 13.2. Newborns gain normal flora through exposure to the environment (vaginal canal, breast-feeding, human contact, doctor's visits, etc.).
9. Moist or oily skin areas have more resident (and transient) flora than does dry skin.
10. Most normal flora are found in the mouth and in the large intestine. Stomach acidity inhibits most bacteria and few normal flora are found in the small intestine. Yogurt has live bacteria that are helpful and prevent colonization by pathogens.
11. The upper respiratory tract harbors normal flora. The lower respiratory tract has no normal flora.
12. Bladder infections are more common in women.
13. True pathogens can cause disease in healthy individuals.
14. Virulence factors are characteristics or structures that enable a pathogen to adhere, multiply, and cause disease in a host.
15. a. capsules—enable adhesion and block phagocytosis, b. Fimbriae—enable adhesion, c. toxins—damage host cells, d. enzymes—break down host extracellular matrix.
16. Biosafety level 1 organisms are not known to cause disease; Biosafety level 4 are extremely dangerous (lethal and untreatable).
17. Entry is important because each pathogen can colonize different areas—having influenza on your hands won't cause disease unless you bring your hands in contact with your respiratory system (i.e., touch your face).
18. a. skin – streptococcus and staphylococcus enter through broken skin, helminths burrow into the skin. b. gastrointestinal tract—food and waterborne pathogens (cholera, *Shigella, Salmonella*) enter through the skin. c. Respiratory—most pathogens that cause respiratory infections (influenza, common cold, tuberculosis) d. urogenital—sexually transmitted diseases.
19. Infectious dose is the minimum number of microbes that must be present for infection to progress. You'd rather drink 1000 cholera cells, since *V. cholera* has a very high infectious dose and measles has an ID of 1.
20. If an organism isn't adherent, it cannot colonize and will become a "transient, " never reproducing or causing disease. See table 13.7.
21. Exotoxins and exoenzymes are both secreted from the cell and cause damage. Exoenzymes break down tissues and extracellular matrix; exotoxins target specific cells. Examples of exoenzymes include mucinase—digests mucous, keratinase—digests keratin, collagenase—digests collagen.
22. Intoxications are due to the ingestion of a preformed toxin, without the pathogen growing within the host. Infections have bacteria or viruses growing in or on a host.
23. See figure 13.8. Tetanus toxin, botulinum toxin, cholera toxin, and diphtheria toxin all contribute greatly to disease processes.
24. See figure 13.13.
25. See figure 13.14. A localized infection stays in a single area, a focal infection starts as a localized infection, then spreads throughout the body (becomes systemic).
26. In a mixed infection, multiple microbes become established at the same time. A primary–secondary infection is seen when a primary infection predisposes a host to develop a second disease.
27. The common cold is acute—it comes quickly and leaves quickly.
28. A sign is an objective (measurable) indication of disease. A symptom is a subjective (nonmeasurable) indication described by the patient. A fever is a sign, a headache is a symptom. A syndrome is a disease with characteristic signs and symptoms.
29. See figure 13.15.
30. Latent infections present in the body, undetected for long periods of time. Herpes, EBV, TB, syphilis, and malaria all have latent stages.
31. Strep throat may lead to rheumatic heart disease, Lyme disease can cause arthritis.
32. A reservoir is the natural habitat for a pathogen; a source is the individual object where the infectious agent was acquired.
33. An asymptomatic carrier is infected, but doesn't have symptoms. A passive carrier is not infected, but carries microbes (see figure 13.16).
34. Mechanical.
35. Biological.
36. No, rabies and West Nile virus are both zooriotic, with non-humans as natural hosts.
37. See figure 13.18.
38. A fomite is an object that transmits pathogens.
39. Tetanus is not communicable because it isn't transmitted directly from person to person.
40. Aerosol spread are fine dust or moisture that contains live pathogens; droplet nuclei are dried bits of mucous from the respiratory tract and carry hardier pathogens.
41. Nosocomial infections are from hospitals.
42. Universal precautions are described by the CDC to limit the spread of disease in a clinical setting. All patient specimens are treated as if they harbor disease (universal).
43. Epidemiology.
44. Diseases that must be reported to public health authorities (CDC).
45. Prevalence is the total number of existing cases. Incidence is the number of new cases in a certain time frame. 20%, 10 cases in 100 for the week.
46. Morbidity rate—number of non-fatal cases of a disease, mortality rate—deaths in an infected population due to a disease.
47. Endemic diseases are normally found in an area.
48. When more cases than expected occur in a given area, an epidemic occurs. The spread of an epidemic across the globe is a pandemic.
49. Most cases in a population go undiagnosed and unreported—what we see is "the tip of the iceberg."

Organizing your Knowledge

Table 1. Distribution of Flora
a. Yes; b. Yes; c. Yes; d. Yes; e. No; f. Yes; g. Yes; h. Internal; i. Internal; j. Internal; k. Internal; l. Internal; m. Internal; n. Internal.

Table 2. Disease Transmission Overview
a.Ingestion; b. Respiratory; c. Generally an STD; d. Respiratory; e. Contaminated wound; f. *Vibrio cholera*, g. HIV; h. *Clostridium tetani;* i. Strep throat; j. Tuberculosis; k. Feces; l. Respiratory; m. Sexually transmitted; n. Respiratory; o. Not a communicable

disease (goes back to environment with decomposition of body).

Self-Test: Vocabulary

1. G	2. N	3. M
4. O	5. I	6. K
7. U	8. Q	9. F
10. V	11. P	12. E
13. J	14. B	15. L
16. A	17. C	18. H
19. S	20. R	

Self-Test: Multiple Choice

1. d	2. c	3. d
4. c	5. b	6. b
7. a	8. d	9. c
10. a	11. c	12. a
13. c	14. a	15. b

Applications of Chapter 13

1. Malaria and other vector-borne diseases such as yellow fever and bubonic diseases are very difficult to track because of the indirect transmission routes—two people can have the same disease in the same town or village, but have no contact with each other if they have contact with the same population of mosquitoes or other vectors.
2. Reportable diseases are all identifiable and most are infectious or virulent enough to pose a risk to the population as a whole.

Chapter 14

Building Your Knowledge

1. See table 14.1. First line of defense—physical, chemical and genetic barriers to infection, second—nonspecific immune cells, third—specific immune cells. The first two are innate and the third is an acquired response that conveys immunity.
2. Physical barriers include the skin, mucous membranes, flushing action of tears, digestive system, and mucosal drainage.
3. The loss of the outer layers of skin helps prevent long-term colonization by pathogens. Skin is also dry and watertight, making it difficult to penetrate.
4. a. Sebaceous secretions; b. High electrolyte concentration makes water unavailable to microbes; c. Direct antimicrobial effects; d. Breaks apart Peptidoglycan; e. Skin; f. Tears and saliva.
5. Genetic defenses between species—several diseases, like chicken pox and mumps are human-only. Genetic defenses within a species—some disorders confer resistance to infectious diseasaes (sickle-cell anemia heterozygotes are healthy and resistant to malaria).
6. a. surveillance of the body; b. recognition of non-self; c. destruction of foreign materials.
7. See figure 14.5. The RES connects the blood and lymphatics to the connective tissue that surrounds cells. It also holds many phagocytes and helps filter body fluids.
8. Hemopoiesis is the production of blood cells. In newborns it takes place in the liver and in adults it takes place in the bone marrow.
9. Granulocytes have tiny granules in them, agranulocytes do not. Both cell types are white blood cells or leukocytes.
10. See figure 14.7. Neutrophils, eosinophils, and basophils are all granulocytes. Neutrophils are phagocytes, eosinophils are antiparasites, and basophils are proallergy and proinflammation. Monocytes and lymphocytes are a granulocytes. Monocytes are active phagocytes, lymphocytes are B cells and T cells and are part of specific immunity.
11. Monocytes leave circulation and develop into macrophages.
12. Dendritic cells have long, thin processes and serve as phagocytes and antigen-presenting cells.
13. Platelets are fragments from megakaryocytes and function to promote blood clotting.
14. Chemotaxis is the movement toward a chemical, diapedesis is the movement toward a chemical through a vessel wall.
15. Primary lymphoid tissues are sites of lymphocyte production and maturation; secondary lymphoid tissues such as lymph nodes concentrate immune cells and "host" microbe-immune interactions.
16. Macrophages and dendritic cells have toll-like receptors (TLRs).
17. TLR recognize molecules commonly found on pathogens, but not found in the host (see table 14.3).
18. a. mobilize and attact immune cells to site of injury; b. begin repairs of tissue damage; c. destroy microbes.
19. a. Rednes;, b. Swelling; c. Dolor; d. Calor; e. Vasodilation, increased blood flow to area; f. Cellular infiltrate (systemic = IL-1 production).
20. Neutrophils are generally the first to arrive by exiting circulation (diapedesis) and entering tissues. Neutrophils are NOT specific to any pathogen, but respond to chemotatic signals.
21. Fevers can be caused by exogenous sources (bacteria, viruses, endotoxin) or endogenous sources (IL-1 or TNF). Fever inhibits cell division in several pathogens, it also limits the amount of iron available in the blood and increases blood flow and effectiveness of phagocytes.
22. Neutrophils, monocytes , macrophages, and dendritic cells are phagocytes that kill bacteria by engulfing the bacteria and fusing the phagosome produced with lysosomal enzymes (see figure14.18).
23. The complement cascade is a system of proteins that circulate in an inactive form but are sequentially activated upon exposure to an immune stimulus. The major steps in the cascade are initiation, amplification, polymerization of C9, and membrane attack. See figure 14.28. This cascade brings leukocytes to a site of inflammation, makes pathogens easier to phagocytose, and may directly kill by attacking membranes.

Organizing Your Knowledge

a. First cells into a site, phagocytes; b. Defense against parasites; c. Direct killing, B cell help; d. Sometimes (not a professional phagocyte, however); e. No; f. Yes; g. No; h, No.

Self-Test: Vocabulary

ACROSS: 1 lymphocytes; 4 neutrophils; 7 hemapoiesis; 9 lysozyme; 11 basophil; 13 leukocytes; 16 diapedesis; 18 tlrs. DOWN: 2 cytokines; 3 interferons; 5 humoral; 6 complement; 8 pyogenic; 10 first; 12 pyrogen; 14 thymus; 15 serum; 17 primary.

Self-Test: Multiple Choice

1. d	2. a	3. b
4. b	5. d	6. c
7. c	8. c	9. b

10. c 11. d 12. b
13. a 14. c 15. d

Applications of Chapter 14

1. Answers will vary. Systemic steroids will prevent adhesion molecules from forming on leukocytes and prevent diapedesis of these cells out of circulation and into tissue spaces. This treatment is effective in many different situations where an overstimulation of the immune response is harmful.
2. The adaptive response gives a longer-lasting immunity and protects over time, however, it takes several days for a B or T cell response to become fully active. During this time, a pathogen could overwhelm the body, killing its host. The innate response, while nonspecific, is a very rapid response.

Chapter 15

Building Your Knowledge

1. Immunocompetence is the ability of the body to react adequately to a wide range of foreign substances or pathogens.
2. B cells and T cells.
3. Specificity and memory.
4. Active immunization exposes the host to antigen; passive immunization give a host premade antibodies. Passive immunization is quicker, but active is longer-lasting.
5. Natural immunity occurs without intervention—such as the immunity you develop once you've had the chickenpox. Artificial immunity is the result of intentional manipulation of the immune response—such as getting your childhood shots.
6. Immunotherapy is a form of artificial PASSIVE immunity, active immunization is a form of artificial ACTIVE immunity.
7. See figure 15.1.
8. B cells mature in the bone marrow and T cells mature in the thymus.
9. Antibody.
10. a. To bind to antigens, b. to recognize self-molecules, c. to receive and transmit chemical messages, d. to aid in cellular development.
11. See figure 15.2.
12. MHC stands for major histocompatibility complex. For drawing see figure 15.3. MHC I molecules are found on all nucleated cells, MHC II are on professional phagocytes (antigen-presenting cells).
13. The clonal selection theory states that a diverse array of B cells and T cells reactive to a wide range of antigens is produced and upon exposure to an antigen, the cells that can react to that antigen proliferate. Lymphocytes do not change to match a given antigen, they have a given antigen specificity and are selected for based on antigen exposure.
14. Self-reactive clones are deleted (both B cells and T cells), so we do not have an autoimmune reaction and that we do develop immune tolerance.
15. Immunoglobulins are large glycoproteins that are produced by B cells and that recognized antigen.
16. See figure 15.6.
17. Both Ig molecules and T cell receptors have variable and constant regions. They also both have Ig-like domains that are stabilized by disulfide bridges. T cell receptors are much smaller than immunoglobulins and are never secreted.
18. In birds, B cells mature in a bursa; in humans they mature in bone marrow with specific stromal cells.
19. B cells go to lymph nodes, the spleen and gut-associated lymphoid tissue. They do not circulate, but find their "home" in particular regions.
20. T cells mature in the thymus.
21. See figure 15.8. Proteins in any form and some capsules make good antigens. Larger molecules make better antigens than smaller ones. A hapten is a molecule that can elicit an immune response, if it is with a carrier molecule.
22. Autoantigens are self-antigens that stimulate an immune response; alloantigens are the basis of blood group and MHC tissue typing and are of concern to transplant specialists.
23. Superantigens are toxins that inappropriately overstimulate a T cell response.
24. Most antigens enter through the respiratory or gastrointestinal systems. Antigens are gathered up and grouped at the lymph nodes.
25. Dendrititc cells, macrophages, and B cells can serve as APCs to helper (CD4+). T cells.
26. See figure 15.10. The MHC II holds antigen and the T cell binds to both antigen and MHC II. It then secretes B cell growth factors and interleukins.
27. No, B cells present antigens on MHC II or MHC I, never on their antibodies.
28. B cells require antigen stimulation, and cytokine exposure. The cytokines come from both macrophages and T cells.
29. After activation a B cell will undergo mitosis (expanding the numbers of reactive cells) and differentiate into memory cells and plasma cells (see figure 15.11).
30. See figure 15.14.
31. See table 15.2. a. IgM; b. IgA; c. Memory response; d. B cell receptors; e. Associated with allergic responses; f. IgG; g. IgA; h. IgD; i. IgE; j. IgM.
32. See figure 15.16.
33. The memory cells produced upon first exposure give a stronger, more sustained antibody response.
34. a. CD4; b. CD8; c. T cell help to B cells.
35. ISG is a form a passive immunization where pooled immunoglobulins are given prophylactically after measles exposure and to replace antibodies in immunosuppressed individuals. It is very short-lived protection.
36. SIG is specific immune globulin—generally taken from hyperimmune donors.
37. See table 15.4
38. The Salk polio vaccine is an inactivated virus. The pathogen is killed or inactivated by radiation, formalin fixation, heat, or other methods. Killed vaccines do not elicit as strong cell-mediated response when compared with attenuated vaccines.
39. The measles, mumps, and rubella (MMR) shot is an attenuated vaccine. Advantages of a live vaccine are that the organisms can multiply but not cause disease, the protection is longer-lasting, fewer doses are required. However, live vaccines need refrigeration, they may be transmitted to other people, and they may revert back to become pathogens.
40. Toxoids are inactivated vaccines. The tetanus shot is a toxoid of the toxin produced by *C. tetani.*

41. a. malaria; b. HIV; c. diarrheal diseases (*E. coli, Shigella*).
42. Genes (encoding antigens from a pathogen) are inserted into a nonpathogen and given to a person as a vaccine. Vaccinia has been used to carry HIV genes, herpes virus genes and mycobacterial genes.
43. DNA vaccines carry the genes from a pathogen, but the person being vaccinated is never exposed to the entire pathogen (in either attenuated or killed form).
44. Herd immunity occurs when enough individuals are vaccinated against a disease so a communicable disease is not transmissible in the population. Herd immunity prevents outbreaks and epidemics of disease.

Organizing Your Knowledge

Table 1. a. All nucleated cells; b. Antigen presenting cells; c. CD8 + T cells; d. CD4+ T cells; e. Inside the cell; f. Things that have been phagocytosed.

Table 2. a. T_{H1}; T_{H2}; b. T_{H1}; T_{H2}; c. T_C; d. T_C; e. T_{H1}; T_{H2};; T_C; f. none; g. T_{H1} and T_{H2}; h. T_C; i. T_C; j. T_{H1}; T_{H2}.

Table 3. a. Both; b. Both; c. Antibody; d. Both; e. Antibody; f. TCR; g. Antibody; h. Antibody; i. TCR; j. Both.

Self-Test Vocabulary

1.	Z	2.	X	3.	H
4.	A	5.	E	6.	Q, W
7.	B	8.	T	9.	R
10.	Y	11.	J	12.	L
13.	F	14.	U	15.	N
16.	O	17.	V	18.	G
19.	M	20.	D		

Self-Test: Multiple Choice

1.	c	2.	b	3.	a
4.	c	5.	b	6.	a
7.	d	8.	d	9.	d
10.	a	11.	a	12.	c
13.	d	14.	d	15.	b

Applications of Chapter 15

1. You would want a passive immunization (antitoxin). If you are trying to eliminate snake venom, you don't need to introduce your immune system to more venom; you need premade antibodies.
2. The macrophage will present viral antigens on MHC I and will be killed by a CD8+ T cell.

Chapter 16

Building Your Knowledge

1. a. immunodeficiency; b. hypersensitivity.
2. See figure 16.1.
3. Both allergies and hypersensitivities are the result of an overactive immune response. However, allergies are immediate reactions, and hypersensitivities are delayed.
4. Both allergies and infections will trigger an immune reaction, so will show similar symptoms.
5. Atopy is a chronic, local reaction; anaphylaxis is an acute, systemic reaction (often fatal).
6. The susceptibility to develop allergies is inherited—not the specific allergy reaction.
7. IgE production.
8. The first exposure is the sensitizing dose which primes an immune response. The provocative dose (second exposure to the same allergen) may lead to anaphylaxis (or milder form of allergy attack).
9. See table 16.2. a. Ingestants; b. Contactants; c. Pollen; dust, dander; d. Bee stings, drugs, hormones.
10. IgE, mast cells, and basophils.
11. Histamine, leukotrienes, prostaglandins.
12. See figure 16.4. Blood vessel dilation (lowered blood pressure), constriction of bronchioles (wheezing), pain, excessive mucus production.
13. Asthma, eczema, and food allergies are all atopic diseases. Asthma is characterized by constriction of the bronchioles, leading to troubled breathing. Eczema is an itchy inflammation of the skin, and food allergies are an inappropriate response to foods and may be mild or severe (life-threatening).
14. Cutaneous anaphylaxis is a localized skin reaction (wheal and flare). Systemic anaphylaxis involves respiratory and circulatory problems and can kill in 15 minutes.
15. Allergies can be diagnosed by skin tests, differential blood counts (high eosinophils or basophils indicates allergies), histamine release tests, and IgE concentrations. The most common of these is the skin test (figure 16.6).
16. a.. antihistamines; b. avoidance of allergen (not always practical); c. desensitization therapy. See figure 16.7.
17. Injections of allergen may stimulate IgG responses, which prevent IgE binding (see figure 16.8).
18. Type II hypersensitivities are mediated by IgG and IgM fixation of complement, not by IgE responses.
19. ABO antigens are sugar residues found on red blood cells. Each blood type A, B, and O is a different type of antigen. O is recessive to A and B, so someone who is AO would have blood type A. Type O is the universal donor because there are no type O antibodies present in the recipient's blood. The universal recipient is type AB, because they don't have anti-A or anti-B antibodies in their serum.
20. If you gave type B blood to someone who is type A, the anti-B antibodies would bind to and lyse with (complement) the donated cells. See figure 16.11.
21. Transfusion reactions are treated with drugs to remove free hemoglobin from the blood.
22. Rh+ and Rh- indicate whether a person's cells have the Rh factor on them or not. Rh- females can be sensitized to Rh+ blood cells when they give birth to a Rh+ child. RhoGAM is an anti-Rh treatment that will prevent the mother's immune response from recognizing Rh+ cells.
23. Type III hypersensitivities are caused by soluble antigen–antibody complexes. They are similar to type I and type II hypersensitivities in that all three are antibody-mediated; however type III hypersensitivities are mediated by soluble immune complexes, not antibodies reacting with cells.
24. Serum sickness is caused by exposure to large doses of foreign antibodies. Symptoms include rashes, kidney problems, fever, joint problems.
25. The Arthus reaction is a localized reaction leading to inflammation at the exposure site (generally an injection site).
26. T cells mediate type IV hypersensitivities in a delayed fashion.
27. T_{H1} cells mediate the tuberculin reaction. A positive test indicates exposure to tuberculosis (or similar mycobacteria).

28. MHC I and MHC II molecules and cytotoxic T cells are responsible for transplant rejection.
29. GVHD may occur when bone marrow cells are donated and attempt to reject their new host. The reaction is the same as rejection of a transplanted organ by a host.
30. Donated tissues may come from self (e.g., autograft of skin), an identical twin (isografts), another individual (allograft), or a different species (xenografts). Autografts and isografts are the most successful.
31. Most autoimmunity cases are type II or type III sensitivities.
32. Autoimmune disorders are more common in females.
33. The sequestered antigen theory states some sites of the body are immune-privileged and if the immune cells "see" them, autoimmunity could occur. The clonal selection theory states autoimmunity is due to self-reactive clones not being deleted during B and T cell development. The immune deficiency theory states that suppressor T cells may be less active than they should be, leading to overactivity of other cells. Last, it has been hypothesized that some cells may inappropriately express MHC II molecules and "prime" an autoimmune response.
34. SLE and RA are both systemic autoimmunities.
35. Diabetes mellitus is an autoimmune reaction causing antibodies to target insulin-producing cells. Multiple sclerosis is an autoimmune disease caused by the destruction of the myelin sheath around nerves.
36. Primary immunodeficiencies are congenital; secondary immunodeficiencies are acquired.
37. Agammaglobulinemia patients have little or no IgG, and are prone to bacterial infections such as *Pseudomonas* or *H. influenzae.*
38. DiGeorge syndrome patients have no thymus (therefore no T cells). They are prone to persistent fungal, protist, and viral infections.
39. A severe combined immunodeficiency may be caused by a lack of lymphocyte stem cells, ADA deficiency, lack of MHC II, or a poorly developed thymus and lymph nodes.
40. a. infection; b. disease; c. chemotherapy; d. radiation.
41. Cancer is a new growth of abnormal cells. Benign tumors do not spread; malignant tumors will spread to other tissues (metastasis).
42. Cytotoxic T cells are able to kill cancer cells in immunoccompetent individuals, but not in AIDS patients, so cancers develop.

Organizing Your Knowledge

Table 1. a. Overactivity; b. Overactivity; c. Underactivity; d. Overactivity.
Table 2. a. Antibodies and cells; b. Cells; c. Cells; d. Antibodies and cells; e. Antibodies; f. IV; g. I; h. III; i. I.
Table 3. a. Different species; b. Identical twin; c. Different person; d. Self; e. Use of pig heart valves; f. Twin donated kidney transplants; g. Cadaver tissues; h. Skin grafts.

Self-Test: Vocabulary

ACROSS: 5 atopy; 7 desensitization; 11 tuberculin; 15 IgE; 17 primary; 18 GVHD; 19 immunodeficiency; 20 xenograft. DOWN: 1 Arthus; 2 asthma; 3 rhogam; 4 hypersensitivity; 6 donor; 8 immunopathology; 9 provocative; 10 allergens; 12 degranulate; 13 anaphylaxis; 14 eczema; 16 histamine.

Self-Test: Multiple Choice

1. a	2. c	3. a
4. b	5. d	6. a
7. c	8. d	9. b
10. b	11. b	12. a
13. d	14. c	15. c

Applications of Chapter 16

1. Desensitization therapy only works with KNOWN allergens. In asthma and eczema, the triggers are often unknown.
2. RhoGAM will not work, since memory B and T cells are already present in the mom and reactive to Rh+ cells, so there will be problems with any Rh+ child in the future. The first child didn't get hemolytic disease because mom wasn't exposed to that child's red blood cells until delivery.
3. The boy is not safe yet. Poison ivy is a delayed type of hypersensitivity, so it takes 2–3 days to fully develop.

Chapter 17

Building Your Knowledge

1. a. phenotypic; b. immunologic, c. genotypic.
2. Staining characteristics (Gram, acid-fast, endospore, capsules, and other structures) can be determined microscopically. Colony morphology descriptions include size, color, speed of growth, and pattern of growth.
3. Biochemical traits are good indicators of microbial metabolism and aid in the identification of genus, species, and strain of bacteria.
4. Genotypic methods do not require culture, can be done with more precision, and are quicker than most phenotypic methods.
5. Viruses are more difficult to culture and more difficult to capture in a sample—it is easier to screen for the presence of antibody than the presence of a virus.
6. Specimen collection is the first step in clinical identification
7. See figure 17.1.
8. Most samples are held in transport containers that have maintenance media, which does not support growth. If samples need to be held, they are refrigerated.
9. Direct tests are known first (Gram stain, immunologic, or genetic methods).
10. Presumptive data place a microbe into a genus or generalized group. Confirmatory data will place a microbe into a particular species or strain.
11. The length of time to analyze a sample varies greatly (minutes to weeks).
12. Serological tests do not require culture of the microbe.
13. Direct antigen testing can be used to identify *Streptococcus pyogenes, Staphylococcus aureus, Haemophilus influenzae*, and other pathogens. This is a rapid (nonculture based) test. False negatives may occur when the microbe is in low concentrations in the sample.
14. Fecal samples have a large number of microbes so isolation media selects for the microbe that is being tested for and against other microbes.
15. A positive biochemical test indicates the sample has a particular enzyme. In clinical labs there are rapid,

miniaturized systems that test several of biochemical traits at the same time in an automated fashion.
16. Phage typing can identify strains of bacteria, which is more specific than simply which genus and species of bacteria.
17. The patient with 10 *M. tuberculosis* is likely sick, since TB are never normal flora. The urine *E. coli* is likely a contaminant from sample collection.
18. Gentotype methods include genetic probes, rRNA analysis, PCR, and G + C base composition.
19. Serology is the in vitro diagnostic testing of serum.
20. Urine, CSF, tissues, and saliva may be tested immunologically.
21. No—you need at least one known variable—either known bacteria and patient samples or unknown bacteria and known antibodies.
22. Specificity is a measure of how specific an antibody is for a given antigen (no cross-reactivity). Sensitivity is a measure of how low the concentration of antigen can be and still be detected. Tests can be sensitive, but may detect more than the target antigen, which means they are not specific. Tests can also be specific, but require a high concentration of antigen in order to react.
23. See figure 17.10. A high titer indicates a greatly dilute sample still is reactive; therefore a lot of antibodies were in the original sample.
24. Agglutination is the clumping of cells; precipitation is the formation of antigen–antibody complexes.
25. Agglutination tests include ABO blood typing, the Widal test for salmonella, RPR test for syphilis, and the cold agglutinin test for mycoplasma.
26. An Ouchterlony is a double diffusion test that will detect unknown antibodies (using known antigens) or unknown antigens (using known antibodies).
27. The Western HIV test tests for more types of antibodies in patient serum and is less likely to be mis-interpreted.
28. See figure 17.14. Hemolysis indicates the patient's serum did not have antibodies in it.
29. Salmonella and pneumococcus may be detected and described by serotyping tests. Antibodies against syphilis may be detected by an immobilization test.
30. See figure 17.15.
31. See figure 17.16. The sandwich ELISA detects antigen in samples. ELISA tests include initial HIV tests, hantavirus, rubella virus, , cholera, rickettsia, and others. Frequently this is a screening test, with a confirmation test following.
32. T cells can be identified by their ability to form rosettes when incubated with red blood cells. B cells can be identified by fluorescent antibody testing.
33. The tuberculin test is a measure of a DTH response and detects exposure to *Mycobacterium tuberculosis.*

Organizing Your Knowledge

Table 1. Sampling Techniques

a. Puncture (venous or arterial); b. Swab; c. Spit; d. Swab (or scrapping then swab); e. Lumbar puncture; f. Swab; g. Clean catch or sterile (catheter); h. Swab.

Table 2. Precipitation vs. Agglutination

a. Small; b. Large; c. Soluble; d. Insoluble; e. Tube precipitation (streptococcal antigens); f. Widal test for salmonella.

Self-Test:Vocabulary

1. I	2. G	3. O
4. B	5. P	6. L
7. K	8. A	9. M
10. C		

Self-Test: Multiple Choice

1. d	2. c	3. a
4. b	5. c	6. b
7. b	8. d	9. c
10. a	11. b	12. c
13. d	14. b	15. b

Applications of Chapter 17

1. Fecal samples are expensive to process, take a long time to culture, and are often inaccurate. Most important, the course of treatment is the same for many different diarrhea-causing organisms, so the culture doesn't add to the clinical care.
2. HIV viremia is very short-lived and most people are asymptomatic (so aren't going to the doctor) when the virus could be initially detected in the blood. The better test is for anti-HIV antibodies, since these antibodies would only appear after exposure to the virus.

Chapter 18

Building Your Knowledge

1. a. *Staphylococcus*; b. *Streptococcus*; c. *Enterococcus*; d.*Neisseria.*
2. *Staphlococcus* is commonly found on the skin and mucous membranes. These bacteria are round and in irregular "grapelike" clusters. They lack flagella and spores and may be encapsulated. *S. aureus* is considered the most serious pathogen.
3. *S. aureus* is very resistant to changes in temperature, pH and salt concentrations. It also resists drying and many disinfectants. This is of concern, especially in institutional settings.
4. See table 18.1.
5. Toxins include hemolysins, leukocidins, enterotoxins, exfoliative toxins, and TSST toxins.
6. Between 20–60% of healthy adults carry *S. aureus*.
7. MRSA is Methicillin resistant *S. aureus* and is spread through contact with skin lesions.
8. A furnuncle is a single hair follicle that is inflamed. Carbuncle is a larger lesion (a cluster of furncles).
9. Osteomyelitis is an infection of the metaphyses of bones (figure 18.4). Pneumonia due to *staphylococcus* is an infection of the lungs. Bacteremia often leads to endocarditis which may progress to death.
10. *S. aureus* has a high salt tolerance and wide temperature range that it grows in, producing toxins which lead to food-borne intoxications.
11. Phagocytes (neutrophils and macrophages) are most protective against *S. aureus* infection.
12. *S. epidermidis* is commonly on the skin and may form biofilms on medical implants, potentially leading to endocarditis if not treated. *S. saprophyticus* may cause urinary tract infections.
13. Staphylococci produce catalase; streptococci do not.
14. *S. aureus* is coagulase positive; the other staphylococci are not (figure 18.6).
15. Catalase is an enzyme that breaks down hydrogen peroxide. Coagulase is an enzyme that coagulates plasma.
16. Staphylococcal infections are treated by clearing away pus and debris from a large area of infection and long-term antibiotic therapy. Penicillin is NOT the drug of choice because most strains are resistant to it.

17. Hand washing, disposal of contaminated material, possible treatment or exclusion of carriers from high-risk areas.
18. Streptococci are cocci that grow in chains. They do not form spores and may have capsules and slime layers. Most streptococci are fastidious, meaning they are difficult to grow in culture.
19. Streptococci are classified by the Lancefield system (serotyping) and hemolytic pattern.
20. Alpha-hemolysis is incomplete—a greenish halo appears on blood agar. Beta-hemolysis is complete—with a zone of clearing appearing around a colony.
21. The viridans streptococci, *S. pneumoniae*, *S. pyogenes*, *S. agalactiae*, and *Enterococcus faecalis* are all potential human pathogens.
22. M proteins resist phagocytosis and improve adhesion.
23. C5a protease eliminates C5a, streptokinase breaks apart clots, hyaluronidase breaks down connective tissue, streptolysins cause hemolysis, pyogenic toxin, and superantigens stimulate T cells.
24. Humans are carriers (about 5–15%) and the only significant reservoir. Skin infections are more common in summer. Throat infections are more common in fall or winter.
25. Pyoderma is also called impetigo and is a skin infection. Streptococcal pharyngitis is an infection of the thoat (strep throat) characterized by purulent discharge on the tonsils. Scarlet fever is caused by pyrogenic toxin-carrying strains of *S. pyogenes*.
26. Rheumatic fever (inflammation of heart, joints, and subcutaneous tissues) and acute glomerulonephritis (kidney disease) are potential long-term complications.
27. A type-specific anti-M-protein antibody is protective against group A streptococcus.
28. *S. agalactiae* is a group B streptococcus and is a common resident of the skin and mucous membranes. Newborns may be at risk for group B streptococcal infection, so women are screened prior to labor and delivery. If they are positive for the bacteria, they are given antibiotics prior to delivery.
29. The enterococci are normally found in the human large intestine. They are a concern because there are multidrug-resistant strains of enterococcus and they are a source of nosocomial infection.
30. Group A streptococcal infections and some group B infections are very serious if not treated properly, which takes proper identification.
31. Rheumatic fever and glomerulonephritis need to be treated as the original infection, because once they have progressed to either condition, there are no specific treatments.
32. These are alpha-hemolytic streptococci that may cause complications with surgical wounds or dental procedures.
33. *S. pneumoniae* causes pneumonia (60–70% of all bacterial pneumonias).
34. Pathogenic strains are smooth—they have a capsule which enables them to adhere and avoid phagocytosis.
35. Most cases of *S. pneumoniae* infection are endogenous. Droplet spread is a possibility; fomite spread is not.
36. Pneumonia, otitis media, meningitis, pneumococcal-infections. There is an effective vaccine that is effective against
37. See figure 18.21. Pink.
38. *Neisseria* are fastidious—they are difficult to culture and are strict parasites.
39. Humans are the only host of this sexually transmitted disease. Symptoms include painful urination, yellowish discharge, pelvic inflammatory disease, and excessive drainage.
40. Meningococcus is *N. meningitditis*, the causative agent of epidemic meningitis. Gonococcus is *N. gonorrhoeae* the causative agent of gonorrhea, and pneumococcus *S. pneumoniae* is the causative agent of most pneumonias.
41. Noncarriers who are in close contact with carriers are most at risk. The disease is spread by droplets or close contact with secretions. Symptoms include fever sore throat, stiff neck, and vomiting. Treatment is generally penicillin G or a cephalosporin. There is also an effective vaccine.

Organizing Your Knowledge

Table 1. a. Breaks down extracellular matrix; b. Kills leukocytes; c. Forms clots, makes it more difficult for leukocytes to reach cells; d. Makes them resistant to penicillin; e. Unknown.

Table 2. a. Metaphyses of long bones; b. Gastrointestinal tract; c. Acne; d. Fever, chills, shock; e. Blisters and exfoliation.

Table 3. a. Positive; b. Catalase +, coagulase +; c. Strep throat, rheumatic fever, impetigo, scarlet fever, AGN; d. Cephalosporins, combined therapy; e. Positive; f. Serotyping, hemolysis patterns; g. Gonorrhea; h. Cephalosporins; i. Negative; j. CSF stained for gram-negative diplococci, rapid tests.

Self-Test: Vocabulary

1. O	2. N	3. P
4. F	5. G	6. C
7. R	8. A	9. M
10. H	11. D	12. Q
13. L	14. I	15. J

Self-Test: Multiple Choice

1. c	2. c	3. c
4. d	5. a	6. b
7. c	8. d	9. b
10. a		

Applications of Chapter 18

1. *Neisseria gonorrhoeae* is a very difficult to grow, fragile microbe that does not survive long in the environment. It is very, very unlikely that anyone could get gonorrhea from a toilet seat.
2. Many strains of gonococcus and pneumococcus are positive for penicillinase, so the first step of treatment for both is to develop a resistance profile for the microbe.

Chapter 19

Building Your Knowledge

1. An endospore is an environmentally resistant, metabolically inert form of bacterial cell. The advantage to the bacteria is the ability to survive starvation or other environmental stressors. Not all gram-positive rods form spores.
2. *Bacillus, Clostridium*, and *Sporolactobacillus* are spore-formers.

3. Cutaneous anthrax is caused by endosopores germinating on small cuts or breaks in the skin—black scar tissue will form. Pulmonary anthrax is due to inhalation and is characterized by respiratory symptoms, cardiovascular shock, and death. Most cases are from livestock in the Middle East, Africa, and Asia. In the United States, some cases have been reported in the textile industry (contaminated wool). The soil microbe, *Bacillus anthracis* causes anthrax. Anthrax is treated with penicillin, tetracycline, and ciprofloxacin.
4. *Clostridium* spp. are strict anaerobes and are catalase negative. *Clostridium perfringens* causes gas gangrene and is a soil microbe. Gas gangrene is prevented by wound cleaning and debridement and treated with hyperbaric oxygen because the organism is an obligate anaerobe that is poisoned by oxygen. Immunization is not possible because the organism has multiple serovars.
5. *C. difficile* causes antibiotic-associated colitis in patients that have recently had broad-spectrum antibiotic therapy.
6. *C. tetani* causes tetanus. As a clostridial species, it is a spore-forming strict anaerobe that is found in the soil. Any wound is capable of being contaminated with this soil microbe, but the puncture wounds are more likely to see germination, since they are out of the oxygen. Tetanus is characterized by spastic paralysis with a fatality rate of 10–70%. Tetanus is prevented with a vaccine and treated with antitoxin, and supportive care.
7. *Clostridium botulinum* causes botulism, which is an intoxication—it is the exposure to the toxin, not the growth of the organism that causes the symptoms. Infant botulism is generally the result of ingesting raw honey. Botulism is prevented by strict food-processing guidelines (D12 standard) and treated with supportive care.
8. Differential diagnoses can be done by determining exoenzyme presence, fermentation patterns, toxin production, and direct ELISA testing.
9. a. *Listeri;*, b. *Erysipelothrix.*
10. Listeriosis is a flulike food-born disease that generally comes from *Listeria monocytogenes*-contaminated dairy products. Immunocompromised individuals and pregnant women specifically are at the greatest risk. *Listeria* is a facultative intracellular parasite, so it moves from one cell directly to another (see figure 19.10).
11. *Erysipelothrix* causes swine erysipelas. People coming into contact with swine tonsils are most at risk—butchers, farmers, vets. Humans will contract erysipeloid, which is treated by penicillin and prevented by wearing gloves or vaccinating pigs.
12. a. *Corynebacterium;* b. *Mycobacteriu;*, c. *Nocardia.*
13. *Corynebacterium diphtheriae* cause diphtheria. It is transmitted by droplet or fomite exposure. Anyone who is unvaccinated and in crowded conditions is at risk. Diphtherotoxin stops protein synthesis and is often the cause of fatalities. The disease is diagnosed by staining and observing the palisade formation of cells (figure 19.12) and serological assays. Treatment includes antitoxin therapy, bed rest, and may include a tracheostomy to allow for unobstructed breathing.
14. The genus *Proprionibacterium* is most associated with acne.
15. Mycobacteria are all acid-fast, non-spore-forming bacilli of irregular shape.Most mycobacteria live in the soil or water. They are acid-fast because of the thick wall of mycolic acid in their cell envelope.
16. Tuberculosis is caused by *Mycobacterium tuberculosis* and is both an ancient and emerging disease. There are mummies with TB and there are cases of multidrug resistant TB in the United States currently. Most people expsosed to and infected by the TB bacillus do not develop symptoms. The treatment for TB is a long course of expensive antibiotics; DOT is directly observed therapy and has worked to limit the development of multidrug-resistant organisms in the United States.
17. Most symptoms are pulmonary, but there can be reactivation TB that may show extrapulmonary symptoms in the lymph nodes, bones, and kidneys. A tubercule is formed when the immune response is unable to clear the bacilli, so it forms a granuloma (see figure 19.17). Tuberculosis is diagnosed by tuberculin testing and confirmatory X ray. Other tests include acid-fast bacteria in sputum.
18. Leprosy is caused by *Mycobacterium leprae*. This disease is endemic in regions of Asia, Africa, Central and South America. Armadillos are used to study leprosy, since no other animal can be infected.
19. See table 19.4.
20. Leprosy can be treated with long-term rifampin and dapsone.
21. *Actinomycetes* may cause actinomycosis, which is an endogenous infection generally due to poor dental hygiene or tooth extraction (see figure 19.26).
22. *Nocardia* species can cause pulmonary, cutaneous, and subcutaneous infections (see figure 19.27).

Organizing Your Knowledge

Table 1. a. Bacillus; b. Bacillus; c. Bacillus; d. Irregular shape; e. Regular shape; f. Gram+; g. Gram +; h. Acid-fast; i. Gram +; j. Gram +; k. Yes; l. Yes; m. Yes; n. No; o. No.

Table 2. a. colitis; b. tuberculosis; c. diphtheria; d. farmers; woolsorters; e. geriatric patients and drug users; f. oeioke with long-term contact with leprontics; g. amputation; hyperbaric oxygen; h. supportive care.

Self-Test: Vocabulary

ACROSS: 1 diphtheria; 4 botulism; 6 debridement; 10 pulmonary; 11 leprosy; 12 listeriosis. DOWN: 2 tubercle; 3 colitis; 5 tetanus; 7 endospore; 8 cutaneous; 9 mycolic.

Self-Test: Multiple Choice

1. c	2. b	3. a
4. c	5. b	6. d
7. d	8. c	9. c
10. c	11. b	12. d
13. a	14. d	15. b

Applications of Chapter 19

1. Culture is NOT the common confirmatory diagnostic test for TB because of the extremely slow growth of *Mycobacterium tuberculosis.*
2. Naturally acquired anthrax is a zoonotic disease and in the past, most of the cases of anthrax were associated with sheep or textile processing.

Chapter 20

Building Your Knowledge

1. They are separated based on oxygen requirements and the ability to ferment lactose.
2. Enteric bacteria are found in the large intestine
3. LPS reacts with the host as an endotoxin. It may cause shock and death. LPS forms the outer membrane of gram-negative bacteria.
4. Pseudomonas is commonly found in soil and water. Diseases include respiratory and skin infections, particularly in cystic fibrosis patients, immunocompromised patients and burn patients.
5. *B. cepia* infections are most commonly seen in cystic fibrosis patients.
6. Malta fever, undulant fever, and Bang disease are all common names for brucellosis. Brucellosis causes abortions in pigs and cows.
7. Humans have a relapsing fever, but do not see the abortions common in animals. Brucellosis is diagnosed by serology and a newer genetics-based test. The treatment is a 3 to 6-week course of tetracycline and rifampin.
8. *Francisella tularensis* causes rabbit fever. This is a zoonotic disease that is endemic in the Northern Hemisphere and is characterized by fevers, chills, headache, and malaise. Later stages of an untreated infection have respiratory and intestinal problems. Treatments include gentamicin and tetracycline.
9. Whooping cough is caused by *Bordetella pertussis*. This bacteria produces a toxin that kills ciliated epithelial cells which leads to mucus buildup and severe coughing to clear the trapped mucus. Recent studies have shown that childhood vaccination may be inadequate to protect for life, which may be why there are more cases now than 20 years ago. Macrolide antibiotics are the treatment of choice.
10. *Legionella* is named for the Legionnaires that first had an outbreak of the disease in 1976. The microbe is normally found in water—and can be found in air conditioners, tap water, pools, etc. The disorder may be diagnosed by symptoms and history of exposure or by an antibody staining test. Treatment includes quinolones and azithromycin.
11. Enteric pathogens frequently cause diarrhea; approximately 4 billion infections with 3 million deaths each year (worldwide).
12. Coliforms, like *E. coli*, ferment lactose rapidly. Noncoliforms, such as *Salmonella,* do not (see figure 20.10).
13. Enrichment media inhibit the growth of nontarget organisms and select for the growth of target organisms. This is important in fecal culture because of the extreme numbers of nontarget organisms.
14. See figure 20.13.
15. It's easiest to grow, so is seen in culture more frequently. It is actually not the most prevalent microbe in the gut.
16. Enterotoxigenic *E. coli* produce toxins that affect the intestine (enterotoxins), enteroinvasive *E. coli* invade tissues, and enteropathogenic *E. coli* cause a wasting diarrhea.
17. Diarrhea is the single greatest cause of infant death in the world. Enterotoxigenic *E.coli* is the most common cause of "travelers diarrhea." Kaolin slows gut motility and may cause greater exposure to the toxin. Pepto-Bismol counteracts the enterotoxin. *E. coli* cause most UTIs as an endogenous overgrowth into the urethra.
18. The other coliforms (*Klebsiella, Serratia, Citrobacte*) are opportunistic pathogens. *Klebsiella* and *Serratia* may cause pneumonia. *Citrobacter* occasionally causes UTIs.
19. *Salmonella* and *Shigella* are primary pathogens and are not found as normal flora in humans.
20. *S. typhi* causes typhoid fever and is transmitted by a fecal–oral route (ingesting food or water contaminated with feces). Typhoid symptoms include fever diarrhea, and abdominal pain. Treatment is antimicrobial drugs or surgical removal of the gall bladder.
21. Enteric fevers are nontyphoid Salmonelloses and are zoonoses. There are approximately 1.5 million cases each year (CDC estimate). *Salmonella* is associated with poultry and reptiles.
22. *Shigella* causes a dysentery called shigellosis. Humans are the only carrier of *Shigella. Shigella* invades the large intestine rather than the small intestine (like S*almonella*). Treatment includes supportive care and antibiotic therapy.
23. Enteric *Yersinia* include *Y. enterocolitica* and *Y. pseudotuberculosis.* These microbes cause severe abdominal pain.
24. *Yersinia pestis* is the causative agent of plague and has capsules, envelope proteins, coagulase, and endotoxin. Plague is endemic in Africa, South America, Asia, the Mideast, and the former USSR.
25. Humans are amplifying hosts, not the natural reservoirs. Pneumonic plague is spread from human to human, bubonic plague is from animal to human. Bubonic plague is called the "black death" because advanced disease states have subcutaneous hemorrhage and gangrene which visibly darken the skin.
26. Plague is treated with streptomycin, tetracycline, and chloramphenicol with a 95% survival rate.
27. a. *H. aegyptius*; b. Respiratory (pneumonia); meningitis; c. Chancroid.

Organizing Your knowledge

a. Pneumonia, septicemia; b. Tularemia; c. Pneumonia; d. Respiratory disease, septicemia, burn complications; e. Dysentery; f. Plague; g. Animals (ruminants); h. Humans (fecal–oral) (some animal transmission); i. Humans, contact; j. Water, airborne (aerosols); k. Food-borne; l. Older vaccinated individuals, unvaccinated; m. Cystic fibrosis patients; n. Small children, overcrowding; o. Cystic fibrosis, burn patients; p. Drinking contaminated water or eating food; q. People in endemic areas.

Self-Test: Vocabulary

1. G	2. I	3. L
4. D	5. Q	6. P
7. J	8. B	9. E
10. M	11. K	12. H
13. A		

Self-Test: Multiple Choice

1. c	2. d	3. c
4. d	5. c	6. c
7. a	8. c	9. b
10. b	11. b	12. b
13. d	14. c	15. a

Applications of Chapter 20

1. Answers will vary. Cystic fibrosis patients are prone to repeat infections with *Pseudomonas* and most are colonized by it to some extent. Most burn victims are not repeatedly exposed and colonized over a long period of time.
2. Answers will vary. Serotyping enables epidemiologists to determine the source of an outbreak and how it moved through the population. In the case of salmonellosis, this is important because there may be multiples ways to come in contact with many different sources of *Salmonella.*

Chapter 21

Building Your Knowledge

1. Spirochetes have a gram-negative cell wall and move with a periplasmic flagella.
2. *Treponema pallidum* causes syphilis. This organism is extremely difficult to grow in lab and extremely fragile in the environment. Most cases of syphilis are transmitted sexually, with as few as 57 organisms causing disease.
3. See table 21.1.
4. Congenital syphilis is contracted by placental transfer of the spirochete to the developing fetus. The early form is from birth to two years with infants showing bone deformation, skin eruptions, and neural damage. The late-form cases have damage to the bones, ears, joints, and teeth.
5. Syphilis is generally diagnosed with dark-field microscopy of a lesion sample. Other tests include rapid plasmin reagin (RPR), immunofluorescence, and the *T. palladium* immobilization test (TPI).
6. Bejel, yaws, and pinta are all treponematoses, but they are not caused by the same subspecies that causes syphilis and are not secually transmitted.
7. *Leptospira interrogans* are also spirochetes. They cause leptospirosis in humans and animals. The spirochetes are transmitted via animal urine contacting human skin (with abrasions. or mucus membranes. It is diagnosed by symptoms and history as well as by a slide agglutination test. It is treated (early) with penicillin or tetracycline. Protective footwear and clothing are the best preventative measures.
8. Borrelia are transited by ticks or lice. *Borrelia hermsii* is the cause of relapsing fever. The disease is characterized by a fluctuating fever, headache, and fatigue. Treatment includes tetracycline and prevention includes improving hygiene.
9. *Borrelia burgdorferi* causes Lyme disease, a disease transmitted by hard ticks, especially in the northeastern United States. The "bull's eye rash" is often seen at the bite site. Late symptoms include cardiac symptoms and chronic fatigue. Treatment includes antibiotic therapy which is more effective early than late. Prevention includes protective clothing and insect repellant.
10. See figure 21.13.
11. Cholera is a water-borne disease characterized by severe water-loss diarrhea. It is spread via fecal–oral routes. Patients die due to rapid dehydration. Most treatments include rapid rehydration (either oral or intravenous).
12. *Vibrio parahaemolyticus* and *Vibrio vulnificus* are both seafood-borne vibrios.
13. *C. jejuni* causes a food-borne gastroenteritis. Symptoms include headache, fever, watery diarhhea. It is diagnosed by symptoms and by dark-field microscopy and treated with rehydration therapy and in serious cases, with antibiotics. *C. fetus* causes abortions in sheep, cattle and goats.
14. *Helicobacter pylori* is the causative agent of stomach ulcers; it is transmitted by fecal–oral or oral routes. Ulcers are treated with acid blockers and antibiotics.
15. Rickettsias include typhus, Rocky Mountain Spotted fever, and ehrlichiosis.
16. Rickettsias are arthropod-borne bacteria (ticks, fleas, lice).
17. *R. prowazekii* is the causative agent of typhus and is transmitted through lice. Symptoms may include high fever, chills, frontal headache, rash, gangrene, and death.
18. RMSF is a rickettsial disease characterized by a spotted rash, fever, chills, and muscle pain and caused by a Rickettsia.
19. *Coxiella burnetii* causes Q fever. Symptoms are similar to RMSF, but *C. burnetti* is an airborne pathogen.
20. *Bartonella henselae* causes cat-scratch disease which can be prevented by rigorous cleaning of scratches.
21. *Chlamydia* species and near relatives cause trachoma, pneumonia, genital chlamydiosis, and ornithosis. Azithromycin works well because these pathogens are obligate intracellular pathogens—any antibiotic used would need to be able to get inside cells.
22. Mycoplasmas lack cell walls and cause STDs and atypical pneumonias ("walking pneumonia"). Since penicillin targets cell walls, it is a poor choice to treat a cell-wall-less bacterial infection.
23. See figure 21.28.
24. *Streptococcus mutans.*
25. Peritonitis is more serious because it is deeper involvement and may lead to tooth loss.
26. Most dental diseases are controlled by preventative dentistry, brushing, and flossing.

Organizing Your Knowledge

Table 1. Disease Overview

a. *Coxiella burnetii;* b. *Treponeme palladium pertenue;* c. *Borrelia burgdorferi;* d. *Helicobacter pylori;* e. *Bartonella henselae;* f. *Rickettsia prowazekii;* g. STD; h. Contaminated animal urine; i. Contaminated water; j. Tick; k. Airborne; l. Contaminated fingers, fomites, flies.

Self-Test: Vocabulary

1. P	2. E	3. Q
4. D	5. B	6. I
7. G	8. A	9. F
10. M	11. H	12. O
13. N	14. K	15. L

Self-Test: Multiple Choice

1. c	2. a	3. d
4. c	5. c	6. b
7. d	8. a	9. b
10. c	11. d	12. b
13. a	14. c	15. b

Applications of Chapter 21

1. Answers will vary. The same behaviors that put people at risk to contract syphilis also put them at risk to contract HIV, gonorrhea, and other STD. The

combination of HIV and any other disease could be deadly.
2. Answers will vary. Most people have been given antibiotics some time for other infections, which will kill the syphilis spirochetes.

Chapter 22

Building Your Knowledge

1. Fungi are ubiquitous; humans are exposed by air, dust, fomites, and endogenous exposure.
2. Most fungi are nonpathogens. Most that can cause disease do so in immunocompromised hosts.
3. See figure 22.1. The yeast form is fast-growing and can be invasive.
4. Most fungi are acquired through environmental exposure. Fungal epidemics occur when there is mass exposure to a common source. A large group of people taking down an old building would all be exposed to histoplasmosis spores.
5. Most pathogens enter through either respiratory or cutaneous exposure or mucosal membranes.
6. a. toxinlike molecules; b. hydrolytic enzymes; c. allergens; d. adhesion factors and capsules.
7. Culturing in lab is very slow and patients don't have that much time before treatment should be started and different fungi have different treatment options. Commonly fungi are diagnosed by KOH mount microscopy, antigen-based tests, or PCR-based tests.
8. See table 22.4.
9. a. plasma membrane; b. fungal cell wall; c. fungal nucleic acid synthesis.
10. See figure 22.7. Most contact is from inhaling spores, particularly around bird feces. The symptoms of infection vary greatly—from asymptomatic to mild cough to chronic histoplasmosis, which resembles TB. The fungal cells evade the immune response by living within macrophages. Histoplasmosis is diagnosed by microscopic observation of samples (sputum, skin, etc.). If necessary, it is treated with Amphotericin B.
11. Valley fever is also called coccidioidomycosis and is caused by *Coccidioides immitis*. Farming, mining, landslides are all associated with the spread of this soil-borne pathogen. Symptoms include coldlike respiratory illness and chronic infection may lead to fungoma formation (figure 22.8). The disorder is diagnosed by microscopic examination of fungal cells in sputum or by serology.
12. Blastomycosis is caused by *Blastomyces dermatitidis* and is endemic to North America and is transmitted by airborne routes, with humans inhaling conidia. The disease has both respiratory (cough, chest pain, fever) and cutaneous (tumorlike scar tissue) symptoms. It is diagnosed throught microscopic observation of yeast cells and treated with Amphotericin B or an azole drug.
13. Rose gardener's disease is also called sporotrichosis and is caused by a common fungal saprobe.
14. No, they are not primary pathogens and are not thermally dimorphic.
15. Dermatophytoses affect the keratinized areas of the body—skin, scalp, hair, nails.
16. See figure 22.5.
17. Ringworm can affect the scalp, beard, body, groin, feet, hands, and nails. It can be diagnosed with a black light or through KOH smear and microscopy. Most cases of ringworm are treated with topical antifungal drugs.
18. Superficial mycoses only affect the upper epidermal layers of the skin and are primarily cosmetic problems. Subcutaneous mycoses are actual infections.
19. *Candida albicans* causes thrush, vulvovaginitis, and systemic infection (rarely). People at risk for these infections are those who have recently taken antibiotics, diabetics, immunocompromised individuals, and pregnant women. Typical treatments are topical antifungal agents (nystatin) or systemic antifungals (fuconazole).
20. *Cryptococus neoformans* can infect the respiratory, nervous and mucocutaneous systems. Bird handlers and immunocompromised individuals are at risk for this opportunistic pathogen. Systemic cryptococcal infections may include meningitis.
21. Cryptococcosis is diagnosed by negative staining of samples to reveal encapsulated yeast cells. It is treated with Amphotericin B and fuconazole.
22. *Pneumocystis (carinii) jirovecci* is an opportunistic pathogen that causes pneumonia. It is spread through droplet or human–human contact. It is treated with pentamidine and cotrimoxazole.
23. Fungal allergens are associated with respiratory problems (e.g., "farmer's lung").

Organizing Your Knowledge

Table 1. a. *Coccidiodes immitis;* b. *Trichophyton, Microsporum, Epidermophyton;* c. *Sporothrix schenckii;* d. histoplasmosis; e. PCP; f. Thrush, VC; g. Cryptococcosis; h. Antigen tests; i. Blacklight, KOH mount scrapings; j. Cigar-shaped yeasts in exudates.

Self-Test: Vocabulary

ACROSS: 3 blastomycosis; 8 dermatophytes; 10 aflatoxin; 11 dimorphism; 12 fluconazole; 13 polyene.
DOWN: 1 sporotrichosis; 2 opportunistic; 4 mycoses; 5 fungoma; 6 respiratory; 7 nystatin; 9 thrush.

Self-Test: Multiple Choice

1. a	2. b	3. c
4. c	5. b	6. c
7. b	8. b	9. b
10. d		

Applications of Chapter 22

1. Answers will vary. Most fungal diseases are self-limiting, with a healthy immune response clearing the infection without the use of antifungals. Also, most antifungals are toxic to the host.
2. Answers will vary. The appearance of HIV/AIDS has opened the door for many more cases of what were formerly rare opportunistic pathogens (PCP, thrush, *Cryptococcus*, etc.).

Chapter 23

Building Your Knowledge

1. Nearly 20% of all infections are caused by parasites.
2. Parasite distribution has increased due to rapid travel, immigration, and an increase in the number of people who are immunocompromised (AIDS).
3. a. sarcodinians; b. ciliates; c. flagellates; d. apicomplexans.
4. Trophozoites are the active, feeding stage while cysts are the more environmentally resistant form that is often the infective body.

5. A karyosome is the large nucleolus found in trophozoites of *Entamoeba histolytica*.
6. Chromatoidals are clustered ribosomes found in the cyst stage of *E. histolytica*.
7. Humans are the primary host for *E. histolytica*. The parasite is common in the tropics, especially where human waste is used as fertilizer. The severity of human infection is dependent upon the strain of the parasite, size or the inoculation, and degree of host resistance.
8. *E. histolytica* penetrates mucosal membranes, causing dysentery, pain, fever, and weight loss. Untreated, the parasite may invade the internal organs and liver abscesses may form. Amebic dysentery is diagnosed by the examination of fecal smears for the presence of the parasite.
9. See figure 23.1. Preventative measures include proper hygiene and sanitation.
10. a. *Naegleria fowleri*; b. *Acanthamoeba.*
11. Both amebic brain infection parasites are found in standing water, especially water that has lots of bacteria in it. Early treatment is crucial because *Naegleria menigoencephalitis* is rapidly fatal.
12. *Balantidium coli* is a ciliate that causes an opportunistic intestinal disease (with nausea and dysentery common). *B. coli* is spread by pigs and contaminated water. Person-to-person transmission is very rare.
13. *Trichomonas, Giardia, Trypanosomes, Leishmania.*
14. Trichomoniasis is a sexually transmitted disease (STD) with inflammation of mucus membranes causing itching, pain, and breakdown of urogenital membranes. The disease is diagnosed by visualizing swimming trichomonads in samples and treated with Flagyl. It is one of the top 3 reported STDs in the United States.
15. *Giardia lamblia* causes giardiasis, a water-borne diarrheal disease. Cysts can remain in water for months. Drinking water from "clean" mountain streams puts hikers and campers at risk. Diagnosis is very difficult since the parasite isn't constantly in the feces of infected patients (intermittent shedding). Preventative measures include hygiene and disinfecting water.
16. They are both disorders where flagellated parasites enter the bloodstream of infected individuals. These parasites are transmitted by blood-sucking insects.
17. See Figure 23.8; amasitgote; tryphomastigote; vector.
18. *T. brucei* causes African sleeping sickness which is spread by tsetse flies. Symptoms include fever, enlarged liver, enlarged spleen, edema, coma, possible death.
19. *T. cruzi* causes Chaga's disease which is spread by reduviid bugs.
20. See figure 23.10.
21. Leishmaniasis is transmitted by sand flies. The promastigote stage is transmitted to the human host, where the parasite matures to the amastigote stage. Cutaneous leishmania infection is a localized infection of skin capillaries. Systemic leishmaniasis is an infection of internal organs and is more serious.
22. a. Plasmodium; b. Toxoplasma; c. Cryptosporidium.
23. *Plasmodium vivax, P. malariae, P. falciparum,* and *P. ovale* cause malaria. 40% of the world's population live in endemic regions with 300–500 million new cases and 2 million deaths each year.
24. Mosquito; human.
25. 2,000 to 40,000 merozoites are produced per infected liver cell. Merozoites are found in the blood. Malaria symptoms include malaise, fever, aches, and relapsing bouts of fever and sweats. The parasites complete sexual reproduction in the mosquito host.
26. Sickle-cell trait changes the shape of red blood cells, making them more difficult to invade. Malaria is treated with chloroquine, or in resistant cases, treated with artemesinin. Control methods include eliminating standing water and using insecticides and bed nets.
27. Vaccine development is hindered by the four different species involved in the disease, multiple stages, and multiple antigenic types in different strains.
28. *T. gondii* is an apicomplexan with nearly universal distribution. The primary reservoirs for the parasite are felines and their prey (rodents). Herbivores eat oocysts, cats eat the herbivores and pseudocysts with them. Pregnant women and AIDS patients are most at risk. The parasite can cross the placenta and cause stillbirth or severe birth defects. The disease can be prevented by using proper hygiene with cat feces and cooking meat fully.
29. *Cryptosporidium* causes gastroenteritis and is a water-borne infection with no satisfactory treatment (it is generally a self-limiting infection, however). It is diagnosed with indirect immunofluorescence-based tests.
30. This disease is transmitted from human to human in a fecal–oral pattern, generally through contaminated food. Symptoms include watery diarrhea, cramps, and weight loss.
31. See table 23.4.
32. Larval development occurs in the intermediate host; mating occurs in the definitive host. Transport hosts do not have a developmental stage, but may be required for completion of the cycle.
33. Humans are exposed by soil, food, water, and infected animals.
34. Adult helminthes don't multiply in the same host.
35. Most helminth diseases occur in rural tropics and subtropics. Most infestations are systemic and symptoms are most severe in populations that are malnourished or in poverty.
36. Intestinal worms often cause intestinal damage leading to malabsorption and a lack of appetite.
37. Eosinoophils are most responsible for eliminating worms.
38. Helminths are diagnosed by patient history, differential blood count with elevated eosinophils, and by visualizing eggs. Most drugs are toxic to the host because helminthes are eucaryotes too—making selective toxicity difficult.
39. Intestinal worms develop inside the intestines; tissue trematodes develop in other soft tissues.
40. *Ascaria* are transmitted by eggs that are released in feces then passed on by food, drink, or object contact. For pathway of infection see figure 23.19.
41. Adult hookworms have oral cutting plates that they use to attach to the intestine. Hookworms aren't ingested—they are able to directly penetrate the skin.
42. Trichinosis is spread through eating undercooked pork. Humans are a dead-end host because no one eats human corpses (burial/cremation and no cannibalism).
43. *Wurcheria bancrofti* causes Bancroftian filariasis and *Onchocerca volvulus* causes river blindness.

44. Symptoms of elephantiasis are severe swelling of the legs and scrotum (or arms) caused by the worms growing in the main lymphatic channels, blocking the lymph's return to circulation and causing marked swelling. See figure 23.23.
45. Black flies transmit the filarial worm that causes river blindness.
46. Schistosome miracidium infect snails. Snails release cercaria which infect humans. Schistosomiasis starts with itching at the point of entry where the cercaria enters the host. This is followed by fever, chills, diarrhea, cough, and enlarged spleen and liver. Schistosomiasis control methods include limiting snail populations in endemic areas.
47. Lung and liver flukes are transmitted by ingestion of undercooked seafood.
48. See figure 23.26.
49. Taeniasis is a tapeworm infestation where eggs or proglottids are eaten by a cow. Humans eat undercooked meat with cysticerci which attach to the intestinal wall and mature to an adult tapeworm that releases eggs and proglottids.
50. There are numerous arthropod-born diseases—parasitic, viral, and bacterial. Commonly the distribution of the vector determines the geographic range of the disease. See table 23.6.

Organizing Your Knowledge

a. Ciliate; b. Amoeba; c. Hemoflagellate; d. Filarial worm; e. Blood fluke; f. Flagellate; g. Filarial worm; h. Cryptosporidiosis (enteric); i. Giardiasis (enteric); j. Meningoencehpalitis; k. Malaria; l. Trichinosis; m. Chaga's disease; n. Pig feces; o. Fecal-oral (water); p. Sand fly; q. Black flies; r. Infected snails (water); s. Sexual contact; t. Mosquito.

Self-Test: Vocabulary

1.	E	2.	O	3.	H
4.	A	5.	N	6.	Q
7.	F	8.	P	9.	C
10.	I	11.	L	12.	M
13.	K	14.	J	15.	D

Self-Test: Multiple Choice

1.	d	2.	b	3.	b
4.	b	5.	d	6.	b
7.	c	8.	b	9.	c
10.	d	11.	d	12.	c
13.	d	14.	c	15.	a

Applications of Chapter 23

1. Answers will vary. River blindness is associated with a bacterial (*Wolbachia*) coinfection with a filarial worm. The antibiotics will kill the bacteria, preventing the inflammatory response that causes blindness.
2. Answers will vary. The sporozoite stage would be better, since one sporozoite infecting one liver cell will release thousands of merozoites. It would be easier to control a few sporozoites than thousands of merozoites.

Chapter 24

Building Your Knowledge

1. Viruses are not cells. They are obligate intracellular parasites and are much smaller than bacteria, fungi, or parasites.
2. Animal viruses are grouped based on nucleic acid (DNA or RNA), capsid type, and presence or absence of an envelope. Most human-pathogen DNA viruses are double-stranded and most RNA viruses are single-stranded. Some viruses have an envelope, which was originally part of the host cell membrane system.
3. Viruses have proteins on their outer surface that interact with specific receptors found on host cells. Hepatitis viruses target the liver, while HIV targets CD4+ T cells.
4. Protective immunity comes from exposure (to the virus or a vaccine) and development of neutralizing antibodies as well as a T cell response.
5. Chronic viral infections have slowly multiplying virus over a long period of time. Latent infections have a cells harboring a dormant virus that is not multiplying. Both are forms of persistent rather than acute infection.
6. Oncogenic viruses are viruses known to cause cancer. Examples of oncogenic viruses include papilloma viruses, Epstein-Barr, and hepatitis B.
7. Prenatal infection with rubella or cytomegalovirus is associated with formation of birth defects.
8. See table 24.1.
9. Poxviruses are large, complex enveloped DNA viruses (see figure 24.1) that reproduce in the cytoplasm of epidermal cells and connective tissues. Viral reproduction at these sites causes abnormal growth and a pox or lesions to form.
10. Variola is the smallpox virus, while vaccinia is the cowpox virus originally used to vaccinate against smallpox.
11. Variola major causes toxemia, intravascular, and shock (death up to 90% of the time in some populations). Smallpox was eradicated from the human population with the last case seen in 1977. Smallpox was transmitted by inhalation of smallpox virions or contact with pox lesions. Symptoms of infection are fever, malaise, and a rash that begins at the pharynx and spreads to the face and extremities. Scarring was frequent.
12. Molluscum contagiosum is a skin disease transmitted by direct contact or fomites. It causes small, smooth waxy lesions to form. Cautery or freezing is the typical treatment.
13. Monkeypox and cowpox can cause disease in humans (see figure 24.4).
14. a. Herpes simplex 1 and 2 (HSV); b. Cytomegalovirus; c. Herpesvirus -6 and -7; d. Chickenpox and shingles; e. Lymphoid tissue infection (mononucleosis); f. Kaposi sarcoma.
15. See table 24.2. Both viruses are human-only (animals can be experimentally infected) and cause lesions and are spread by direct contact. The virus can become latent by entering the trigeminal nerve (HSV-1) or sacral ganglia (HSV-2). Recurrence is associated with fever, stress, mechanical injury. During a recurrent attack the virus begins actively replicating again. HSV-1 in children is most commonly seen as fever blisters; HSV-2 is a STD whose symptoms include urethritis, itching cervicitis, and lesion formation. Antibody tests are confirmatory for HSV infection, but the presence of giant cells or eosinophilic inclusion bodies is indicative of HSV infection. Treatments reduce viral shedding and length of outbreaks (Acyclovir and valacyclovir). Whitlows are local infections (see figure 24.10) with herpes, usually due to occupational exposure in health care workers.

16. Varicella-zoster virus (VZV) causes chickenpox and shingles. Shingles is a reactivation of the virus that caused chickenpox years earlier. Yes, you can catch chickenpox from someone with shingles. Lesions of shingles follow the innervation pattern from the nerve they were latent in (see figure 24.12). Reactivation of the virus and the appearance of shingles is often seen with declining immune responses or stress. There is a vaccine currently available that prevents chickenpox.
17. CMV is cytomegalovirus, a herpesvirus with widespread distribution in the human population. Fetuses, newborns, and immunodeficient adults are at risk of serious CMV complications. Congential CMV symptoms include an enlarged liver and spleen, jaundice, microcephaly, ocular inflammation, and death. AIDS patients are at risk for disseminated CMV, a systemic disease with multiple organ failure and a high mortality rate. Vaccines have not been developed because there is no animal model and reinfection is common, so it is uncertain whether vaccination would be protective. Also, there is concern about the oncogenic potential of an attenuated virus.
18. Both diseases are caused by the Epstein-Barr virus.
19. In developing countries, early exposure leads to more cases of Burkitt lymphoma. People living in industrialized countries are exposed later in life and tend to develop mononucleosis.
20. a. Infectious mononucleosis; b. Burkitt lymphoma; c. nasopharyngeal carcinoma.
21. HBV-6 causes roseola in infants, a mononucleosis-like disease in adults, and may be correlated with MS.
22. Seropositive status does not indicate causation, since the herpesviruses have a widespread distribution and prior exposure any time would render someone sero-positive.
23. Hepadnaviruses are the major group of viruses that cause hepatitis.
24. Three major viruses cause hepatitis (see table 24.3). They are not related to one another (two are RNA viruses and one is a DNA virus).
25. Hepatitis B has a very low infectious dose and is spread by close contact or through blood or serum products. Symptoms include rash, lowered white blood cell count, abdominal pain, fever, and arthritis. There is also a connection to hepatocellular carcinoma, a form of liver cancer. Common preventative measures are safer sex practices, universal precautions when handling blood products, and vaccination.
26. Adenoviruses cause respiratory symptoms—they are some of the cold viruses. They can also cause eye infections and acute cystitis. These viruses are spread by close contact and secretions.
27. Papillomaviruses cause wart formation (common, plantar, genital warts) and are associated with cervical cancer. There is currently a HPV vaccine that has been shown to reduce the risk of HPV and therefore cancer.
28. Parvoviruses are the only human-infecting single-stranded DNA viruses. They cause fifth disease in children, fetal severe anemia, distemper in cats, and cardiac infection in puppies.

Organizing Your Knowledge

a. Enveloped double-stranded DNA; b. Enveloped double-stranded DNA; c. Enveloped double-stranded DNA; d. Naked single-stranded DNA; e. Enveloped double-stranded DNA; f. Common cold; g. Mononucleosis, Burkitt lymphoma; h. Herpes simplex I; i. Warts, cervical cancer; j. Chickenpox and shingles; k. Ganciclovir (for immunosuppressed; l. Vaccination (passive and active); m. Avoid contact; n. Avoid contact (especially pregnant women); o. Vaccine (vaccinia virus)

Self-Test: Vocabulary

1.	N	2.	B	3.	P
4.	I	5.	F	6.	L
7.	A	8.	Q	9.	C
10.	G	11.	D	12.	K
13.	O	14.	J	15.	E

Self-Test: Multiple Choice

1.	b	2.	b	3.	a
4.	d	5.	d	6.	c
7.	d	8.	a	9.	b
10.	b	11.	c	12.	b
13.	d	14.	a	15.	d

Applications of Chapter 24

1. Viruses are obligate intracellular parasites, so there is very little that can be done to damage a virus that won't also damage the host. Antiviral agents are generally very toxic, therefore the best course of action is to develop preventative measures, such as vaccines.
2. Most DNA viruses have a wide distribution and cancer has many different causes. The chance of someone coincidentally having a virus and a cancer is very high, but correlation does not indicate causation.

Chapter 25

Building Your Knowledge

1. There are 12 groups of RNA viruses that infect humans. They are separated based on the presence or absence of an envelope and nature of their RNA genome. See table 25.1
2. See figure 25.1.
3. Hemagglutinin is needed to bind to mucosal receptors and for the viral particle to enter host cells. Neuraminidase is an enzyme that breaks down respiratory mucus and assists in fusion of virus to host cell.
4. Antigenic shift occurs when with coinfection the RNA from one influenza virus is traded for that of another influenza virus (see figure 25.3). Antigenic drift is the accumulation of mutations in glycoprotein genes. Antigenic shift would not be possible if the RNA genome wasn't already in pieces (i.e., segmented). Influenza has both antigenic drift and shift, so the flu viruses are different each year.
5. Influenza viruses are named for the virus type (A, B, C), animal of origin, location, and year of origin. The influenza A are also named for the hemagglutinin and neuraminidase (H1N3, etc.). The 1918 influenza pandemic killed more people than died in WWI.
6. The anti-influenza drugs inhibit endocytosis, uncoating, and budding and include amatidine, rimantidine, zanamirvir, and oseltamivir.
7. Virus variants are chosen, the CDC provides the correct viral stocks to vaccine manufacturers. Viral stocks are inoculated in eggs, incubated, and harvested. The harvested virus is chemically inactivated and checked for purity and sterility. The vaccine is then made available for use.
8. Flu-mist is an attenuated (cold-adapted) live virus.

9. Hantavirus causes a pulmonary syndrome with high fever, lung edema, and pulmonary failure. The disease is spread by rodents.
10. a. Paramyxovirus; b. Pneumovirus; c. Measles; all three are spread by respiratory droplet.
11. Parainfluenza causes a mild upper respiratory syndrome, with possible bronchitis or croup developing, especially in young children who have not developed immunity.
12. Mumps is transmitted through saliva and respiratory secretions. Early in an infection, the virus replicates in salivary glands. Mumps is diagnosed based on viral antigens and treated by alleviating symptoms. The disease is prevented with the MMR vaccine.
13. Humans are the only known hosts for measles.
14. Measles is transmitted by respiratory aerosol contact. People are infectious from incubation to skin rash, but not during recovery. Measles is diagnosed by clinical observation (Koplik's spots and history of exposure). Treatment is supportive care and prevention is achieved by the MMR vaccine.
15. RSV in adults is a fairly mild respiratory illness. In infants, especially medically fragile infants, the disease causes wheezing, rales, and difficulty breathing.
16. Rabies is a rhabdovirus with a nearly 100% mortality if left untreated. It is a zoonotic disorder. The virus multiplies at the injury site at first, then it moves up the nervous system, eventually reaching the brain. In furious rabies, patients are agitated, disoriented, and may have seizures. In the dumb form of rabies, patients are paralyzed and disoriented. The course of treatment for a diagnosed case of rabies is to administer antibodies (human rabies immune globulin) and then vaccinate the patient with an inactivated vaccine. This is unusual in that a vaccine given AFTER exposure is protective.
17. Coronaviruses are large viruses with distinctive spikes on their envelope. SARS is caused by a coronavirus.
18. Togaviruses are RNA viruses with a loose envelope (like a toga, hence the name).
19. Rubella is German measles and is spread by contact with respiratory secretions. Congenital rubella may cause miscarriage or multiple birth defects. Postnatal rubella is a mild disease with a mild rash and some joint inflammation. Rubella infection is diagnosed by serological testing and the disease is prevented by the MMR vaccine.
20. Arboviruses are spread by arthropod vectors.
21. The vector's life cycle will determine the distribution and frequency of vector-borne infections. More infections are seen in conditions that favor the vector's survival.
22. The most common viral encephalitis causes are eastern equine, St. Louis, and California encephalitis. Humans are an accidental host.
23. Yellow fever and Denuge fever are both arboviruses carried by mosquitos and cause hemorrhagic fevers.
24. Retroviruses have reverse transcriptase which can convert RNA messages into DNA.
25. See figure 18.13.
26. HIV blood screening tests for the presences of anti-HIV antibodies in a patient's serum.
27. See figure 25.14.
28. See figure 25.16.
29. AIDS-defining illnesses are opportunistic pathogens such as CMV retinitis, fungal infections, herpes-simplex bronchitis or unusual cancers (Kaposi's sarcoma). See table 25.1A for a complete list.
30. Adequate testing has enabled doctors to identify and treat pregnant women with anti-HIV drugs that limit the spread of the virus from mother to fetus.
31. HIV is a sexually transmitted disease that can also be transmitted by contact with contaminated blood or blood products. Fomites are NOT part of the infection cycle.
32. Both infected B and T cells will undergo programmed cell death. HIV particles may enter the brain when infected macrophages cross the blood–brain barrier, releasing viral particles into the brain.
33. There may be a false-negative if a person is tested during the lag period when a person is infected (and infectious), but does not have antibodies in circulation.
34. Reverse transcriptase inhibitors block the conversion of RNA to DNA. Protease inhibitors cause abnormal viral particles to be released. Fusion inhibitors block viral fusion with target cell membranes.
35. HTLV-I is a retrovirus that causes adult T cell leukemia and HTLV-II causes hairy-cell leukemia.
36. Polio is caused by a picornavirus—a small naked, single-stranded RNA virus.
37. Polio is transmitted by a fecal–oral route. Viral particles attach to the oropharnyx and intestinal mucosa where they replicate rapidly, leading to shedding of virus in feces and a viremia. The viral particles move to the nervous system where they can cause a range of symptoms (aseptic meningitis, paralysis, or mild illness). See figures 25.23.
38. The Salk vaccine is an inactivated virus and the Sabin vaccine is an attenuated (live) virus. While the live vaccine gives a longer-lasting immunity, there is a chance that it will revert to a pathogenic form. The Sabin vaccine is no longer routinely given in the United States.
39. Hepatitis A is a nonpolio enterovirus that causes a short-term, infectious hepatitis. Hepatitis A and hepatitis B viruses are not related, but do both target the liver.
40. Human rhinoviruses are the major cause of the common cold. These viruses grow at temperatures below normal human body temperature, which is why they are restricted to the upper respiratory system. Since fomite spread is common, the single best way to limit the spread of rhinoviruses is frequent hand washing.
41. Both Norwalk virus and rotavirus cause gastroenteritis. However, Norwalk virus is a single-stranded RNA virus and rotavirus is a double-stranded RNA virus. Small children are particularly prone to rotavirus infection because they have poor hygiene and have not developed immunity to the virus. These diseases are treated with rehydration therapy (oral or intravenous).
42. Prions are infectious proteins—they lack nucleic acid. Prions are known to cause transmissible spongiform encephalopathies, such as scrapie in sheep, mad cow disease in cattle, and kuru in humans. A variant of Creutzfeldt-Jakob disease is also transmissible and prion-linked. Prions are abnormally folded proteins that serve as templates for cellular proteins to misfold

with. Clumps of these misfolded proteins will kill nervous tissue.

Organizing your Knowledge

a. Enterovirus; b. Orthomyxovirus; c. Paramyxovirus; d. Rhabdovirus; e. Penumovirus; f. Arbovirus; g. Fomites/droplet; h. Sexual contact, contact with blood products; i. Aerosol; j. Fecal/oral; k. Fecal/oral; l. Aerosol.

Self-Test: Vocabulary

ACROSS: 3 dyspnea; 5 rubella; 7 arboviruses; 10 retrovirus; 12 croup; 13 shift; 14 prion; 15 rabies; 16 hantavirus. DOWN: 1 mumps; 2 neurotropic; 3 drift; 4 hemagglutinin; 6 measles; 8 rhinovirus; 9 rotavirus; 11 sabin.

Self-Test: Multiple Choice

1.	d	2.	a	3.	c
4.	b	5.	a	6.	d
7.	d	8.	b	9.	c
10.	c	11.	d	12.	c
13.	d	14.	b	15.	b

Applications of Chapter 25

1. One of the key steps in the development of the flu vaccine is the growth of the virus in eggs to produce large quantities of virus for use in vaccines. Avian influenza kills the eggs, which does not allow for the amplification step to take place.
2. Answers will vary. The chances of a pandemic flu are low, since multiple random events would have to take place to cause the virus to maintain virulence AND be transmitted from person to person.
3. The common cold has literally hundreds of different viruses, each with different strains as its causative agent. The variability in antigens would make it almost impossible to develop a single vaccine for all the different viruses.

Chapter 26

Building Your Knowledge

1. Environmental microbiology is the study of microbes in their natural habitats; applied microbiology is the study of practical uses for microbes and microbial processes.
2. See figure 26.2.
3. A community is all life in an area; a population is a group of organisms of the same species in a given area.
4. A scavenger would likely have a broader niche, since it has a wide range of food sources and is not restricted in atmosphere (nitrogen fixers are anaerobes).
5. Producers are autotrophic and produce organic carbon from carbon dioxide, generally by photosynthesis.
6. Without decomposers, matter would be trapped in organic states and cycling between organic and inorganic forms would stop.
7. The energy available decreases dramatically (approximately 90% is lost at each level).
8. a. all nutrients come from a long-term inorganic reservoir; b. elements cycle between organic and inorganic forms; c. recycling of matter maintains a balance; d. cycles rely on interplay between producers, consumers and decomposers; e. all organisms participate in recycling, however, only certain organisms are "fixers."
9. Geomicrobiology is the study of biological involvement in geological processes.
10. The carbon cycle is most closely associated with living systems.
11. Photosynthesis is the most common form of carbon fixation. Carbon dioxide is released through respiration and decomposition. Methane is released by methanogens.
12. Photosynthetic pigments (chlorophylls, carotenioids, phycobilins) gather light at different wavelengths. The Calvin cycle (light independent) uses the energy gathered from the light dependent reactions to fix carbon dioxide.
13. ATP is synthesized by photophosphorylation, which uses a proton motive force to convert ADP + P to ATP. (See figure 26.8).
14. Anoxygenic photosynthesis uses sulfur compounds (H2S) instead of water as a source of electrons. Sulfur (S2) is generated rather than oxygen. Plants and cyanobacteria are oxygenic photosynthesizers.
15. Bacteria that live in nodules—the rhizobia—fix nitrogen. The legumes that have nodules don't fix the nitrogen themselves, they provide an environment for nitrogen fixation to take place.
16. *Nitromonas* and *Nitrococcus* are nitrifying bacteria. Bacillus and *Pseudomonas* are some of the denitrifying bacteria.
17. Sulfur is used in biotin, cysteine, and methionine.
18. *Thiobacillus* are chemolithotrophs that oxidize elemental sulfur for energy. They produce sulfuric acid in the process, which relases phosphate from rocks.
19. Phosphorus is used in nucleic acids and phospholipids. Phosphate is in fertilizers because it is often the limiting nutrient in terrestrial systems. Excessive phosphorus causes an overgrowth of algae in the hydrosphere (eutrophication).
20. Bioamplification is the concentration of pollutants through a trophic structure.
21. Humus is the slowly decaying organic matter in the soil. Warm soils produce a lot of humus, however, in the tropics this is used up quickly to support plant growth. In bogs, there is a slow rate of decomposition so nutrients are trapped in an organic state. This makes bog soils and jungle soil nutrient poor, but for different reasons.
22. The rhizosphere is the zone of soil surrounding roots, generally containing bacteria, fungi, and protists.
23. Mycorrhizae are fungi in close association with plant roots that help capture water and minerals (and nematodes) for the plants' use.
24. See figure 26.17.
25. Aquifers are being depleted faster than they are being replenished and they are collecting pollutants.
26. See figure 26.18.
27. Thermal stratification is the separation of cool water and warm water regions in a lake or large body of water. These regions are separated by a thermocline.
28. Bacteriophage are a natural control mechanism to keep bacterial counts under control.
29. Human water-borne pathogens include *Vibrios, Salmonella, Shigella, Cryptosporidium,* and *Giardia.*
30. Coliform bacteria survive but do not replicate in water. They are easy to culture. Bacteriphage will

replicate as long as there are bacteria and they are more difficult to culture.

31. Standard plate counts simply sample (generally with dilution, possibly with a MPN system) water. Membrane filtration systems filter water and incubate the filter, looking for bacteria in very dilute samples.

32. Zero fecal coliforms are allowed in drinking water.

33. See figure26.23.

34. See figure 26.24.

35. Fermentation is a controlled mass culture set up to produce organic compounds.

36. Microbes are of benefit to food production when they are used as fermenting agents (e.g., yeasts to make beer, rennin to make cheese). Detrimental actions are food-spoilage organisms that change the food to make it unpalatable and food-borne pathogens, which cause disease.

37. Spoilage organisms do not cause disease, so spoiled food can be safe to eat. Disease organisms don't generally spoil food, so food that can cause disease is generally palatable.

38. Both processes rely on yeast fermentation to produce carbon dioxide (bubbles in beer and bread that raises). Alcohol is also produced, but evaporates in the bread baking process.

39. See figure 26.26.

40. The alcohol will inhibit the growth of yeast.

41. a. Grain mash; b. Wine; c. Vodka; d. Bourbon.

42. The process of cheese making is the fermentation of milk and removal of waste products (whey). See figure 26. 30. Rennin is casein coagulase which is an enzyme isolated from the stomach of an unweaned calf. This enzyme catalyzes the curdling reaction.

43. Food-borne infections differ from intoxications in that with infection the microbes actually grow in the host body. Intoxications are due to toxin exposure (see figure 26.31).

44. There are an estimated 76 million food-borne illnesses each year.

45. UV radiation does not penetrate. Gamma radiation would penetrate the food. Irradiated food does NOT become radioactive.

46. Primary metabolites are essential to the functioning of the microbe. Secondary metabolites are not critical to the life processes of the microbe.

47. Yields can be increased by using a fermentor in continuous culture mode, where bacteria are maintained in log phase (their healthiest state). Also, microbes are selected for their ability to produce the desired end products.

48. See figure 26.38.

49. Continuous systems maintain bacterial cultures in log phase, batch cultures would mature to stationary phase as toxins build up and nutrients are depleted.

50. Penicillin.

Organizing Your Knowledge

Applied microbiology: a., b., c., i., j., k., l.

Environmental Microbiology: a., b., d., e., f., g., h., m.

Self-Test: Vocabulary

1. O	2. J	3. C
4. A	5. M	6. E
7. I	8. K	9. F
10. B	11. P	12. G
13. H	14. L	15. D

Self-Test: Multiple Choice

1. b	2. d	3. d
4. a	5. b	6. d
7. d	8. b	9. b
10. b	11. d	12. b
13. a	14. c	15. c

Applications of Chapter 26

1. Answers will vary. The greatest advantage is that theoretically microbes can be taken to the site of pollution and work continuously without additional energy being added.

2. Answers will vary. Biomining microbes cannot be harmed by the heavy metal they are metabolizing and should be able to work at high-temperature and acidic conditions.

Notes